Angiogenesis

Progress in Applied Microcirculation
Mikrozirkulation in Forschung und Klinik

Vol. 4

Series Editors
K. Messmer, Heidelberg; *F. Hammersen*, Munich

KARGER

S. Karger · Basel · München · Paris · London · New York · Tokyo · Sydney

Proceedings of the British Microcirculation Society, The Medical School, University of Birmingham, April 12–13, 1983

Angiogenesis

Volume Editors
F. Hammersen, Munich; *O. Hudlicka*, Birmingham

38 figures and 9 tables, 1984

KARGER

S. Karger · Basel · München · Paris · London · New York · Tokyo · Sydney

Progress in Applied Microcirculation
Mikrozirkulation in Forschung und Klinik

National Library of Medicine, Cataloging in Publication
British Microcirculation Society. Scientific Meeting (1983 : University of Birmingham)
Angiogenesis : Proceedings of the British Microcirculation Society, Scientific Meeting & Annual General Meeting, Department of Physiology, The Medical School, University of Birmingham, Birmingham, April 12/13, 1983 / volume editors: Hammersen/Hudlicka – Basel ; New York : Karger, 1984.
(Progress in applied microcirculation = Mikrozirkulation in Forschung und Klinik ; v. 4) Includes index.
1. Capillaries – growth & development – congresses
2. Growth Substances – congresses I. Hammersen, Frithjof II. British Microcirculation Society. General Meeting (1983 : University of Birmingham) III. Title IV. Series: Progress in applied microcirculation ; v. 4
ISBN 3-8055-3883-9
W1 PR666FI v. 4
[WG 700 B862a 1983]

Printed in Germany by Mühlberger KG, D-8900 Augsburg
ISBN 3-8055-3883-9

Contents

Preface

This volume summarizes most of the material presented at the meeting of the "British Microcirculation Society on Angiogenesis" held in Birmingham in April 1983 ("Growth of Capillaries in Skeletal Muscles" is discussed in the "3rd Bodensee Symposium on Microcirculation", Prog. appl. Microcirc., vol. 5). There are two main limitations on different aspects of angiogenesis discussed in this volume: firstly, an one-day meeting obviously cannot deal with the numerous problems of this topic, and secondly, the majority of the contributors were from Great Britain. Even so, the material points out the importance of vessel growth in normal tissues (skeletal muscle, heart, retina) as well as in tumours. It discusses some of the problems connected with the isolation and testing of angiogenic factors and the different characteristics of endothelial cells in large vessels and capillaries in normal tissues on one hand and in tumours on the other hand. Furthermore, it indicates that different angiogenic factors cannot be wholly responsible for capillary growth, particularly in normal muscle and heart, and that more attention should be paid to different factors connected with the changes in the microcirculation. As usual, the papers presented, raised more questions than they answered. We hope that this volume will at least give a clear picture of the state of research in several laboratories, and perhaps initiates further work.

Munich/Birmingham, 1984

Frithjof Hammersen
Olga Hudlicka

Prog. appl. Microcirc., vol. 4, pp. 1–8 (Karger, Basel 1984)

Growth of Vessels – Historical Review

O. Hudlicka

Department of Physiology, University of Birmingham, Birmingham, England

Growth of blood vessels is the centre of attention of many scientists from different fields and different points of view: stimulation of capillary growth can be beneficial in the heart in cases of pathological cardiac hypertrophy due to hypertension or valvular lesions or for formation of collateral circulation in cases of myocardial infarction. Prevention or limitation of vessel growth in malignant tumours can limit tumour growth and possibly prevent the development of secondary growth. Growth of vessels is also important in wound healing and regeneration of different tissues, and, of course, in the course of development, and these were actually the conditions when it was first described. *John Hunter* [13] in his "Treatise on the blood, inflammation and gunshot wounds" published in London in 1794, wrote about the development of vessels in the chick embryo: "Blood is formed before the vessels, and when coagulated, the vessels appear to arise. When new vessels are produced in part they are not always elongations from the original ones, but newly formed". *Hunter* also described growth of vessels in inflammation where he again assumed the growth to start from coagulated blood. He saw increased vascularization of tumours and observed enlargement of carotid arteries in deers at the time of vascular growth in newly formed antlers. As we know now, his ideas of formation of new vessels from coagulated blood were not right, but he came close to the concept of the development of both blood and endothelial cells from a primitive mesoderm described much later by *Sabin* [22]:

primitive mesoderm → mesenchymal cells → angioblasts → blood islands → erythroblasts; angioblasts → endothelium

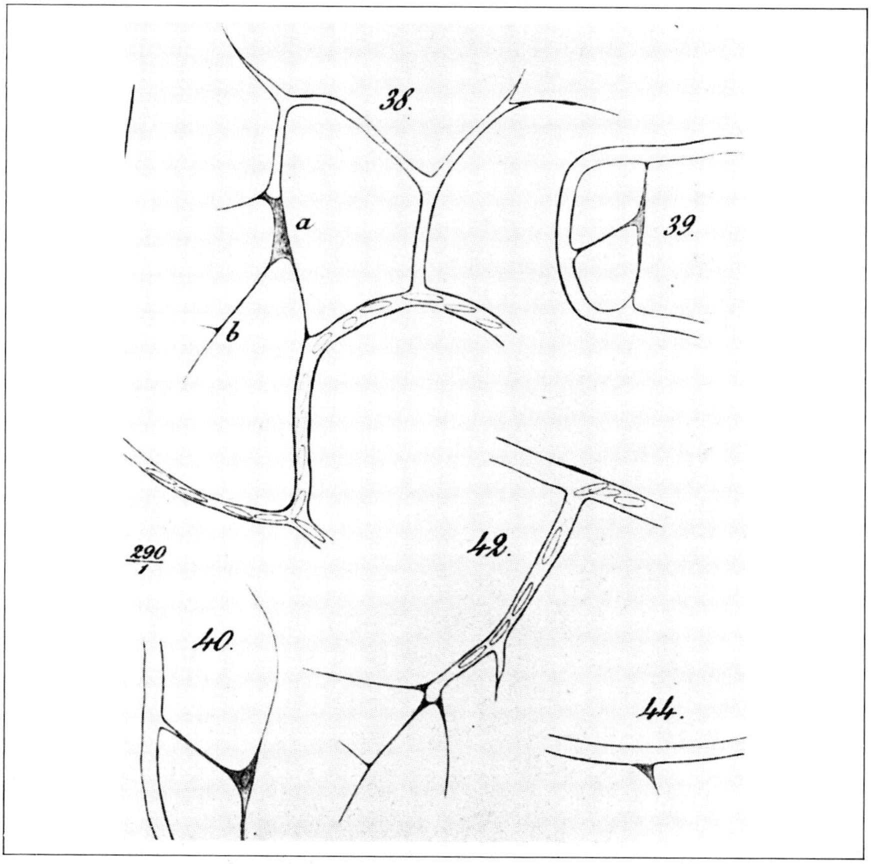

Fig. 1. Reproduction of several drawings of capillary sprouts from a tadpole tail. From *Meyer* [18].

Meyer [18] described correctly growth of new capillaries from preexisting vessels as "thin fibres formed from spindle-like cells which later became hollow" – i. e. capillary sprouts (fig. 1). These were observed even earlier, in 1844 by *Platner* [20] in tadpole tails, but *Meyer* and later *Billroth* [1] extended the observations on vessel growth by sprouting to healing wounds and tumours.

The question what factors were responsible for vessel growth was a matter of controversy from a very early stage of these investigations. *Landerer* [15] suggested the importance of mechanical factors connected with changes in blood flow: he saw formation of sprouts in the places of smallest resistance in the vascular beds. *Thoma* [23, 24] presented evidence that a combination of velocity of flow and/or pressure and growth of surrounding tissue and resulting wall stress are factors initiating vascular sprouting and growth in chick embryos. However, *Loeb* [17] found that development of circulation proceeded in fish embryos even after the heart beat had been stopped by addition of KCl or alcohol to the water in which they were kept, and *Knower* [14] and *Chapman* [4] made similar observations in frog and chick embryos. So other than mechanical factors – possibly some connected with hypoxia or accumulation of waste products – have been postulated as important stimuli for vessel growth.

There is some other evidence supporting the hypothesis that mechanical factors connected with increased capillary flow or pressure are important in the growth of new vessels. *Clark* [5] observed capillary sprouting in tadpole tails preferentially from capillaries with high flow. Capillaries with slow flow eventually regressed (fig. 2). *Clark and Clark* [6] made similar observations in rabbit ears using an implanted ear chamber. They showed in a further study [7] that capillary sprouts had adventitial cells apposed to them, and as the flow increased the newly formed capillary eventually became a venule with more adventitial cells (fig. 3). Possibly other mechanical factors than flow or pressure can be involved in capillary growth. *Brånemark* [3] suggested that oscillatory movements of erythrocytes are important in the "opening up" of newly formed capillary sprouts. Such movements were seen in capillary sprouts in muscles subjected to long-term electrical stimulation ([19], fig. 4). In these experiments sprouts always started to grow from a bending of an existing capillary, and it might be possible that shear stress which is known to stimulate endothelial proliferation in large vessels [11] can play some role in the stimulation of capillary growth.

The mechanical factors are probably important in growth of vessels during development and in normal tissues. On the other hand, formation of vessels in inflammation, wound healing and in tumours, and surely under many other conditions, is dependent on different metabolic factors derived from different blood components, mast cells and tumour cells. It is interesting to know that *Boerhaave* in 1747 [2] commented on the presence of pus aiding the connection of different vessels in wound healing.

The tremendous development in the techniques of culturing endothelial cells in vitro in the last two decades enabled us to study the many different factors under more defined conditions, and to show how capillaries can be formed in tissue cultures when factors such as blood flow are

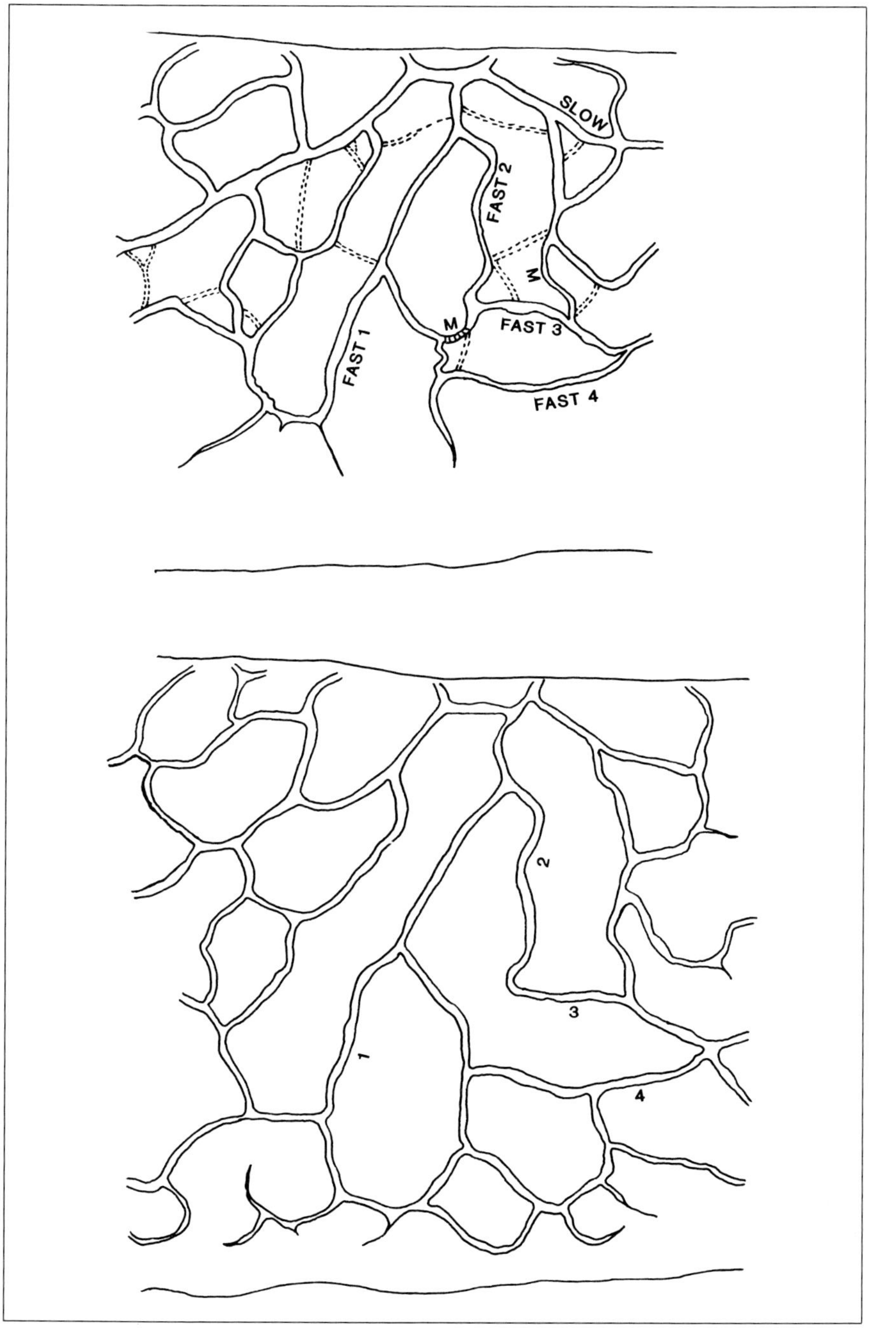

2

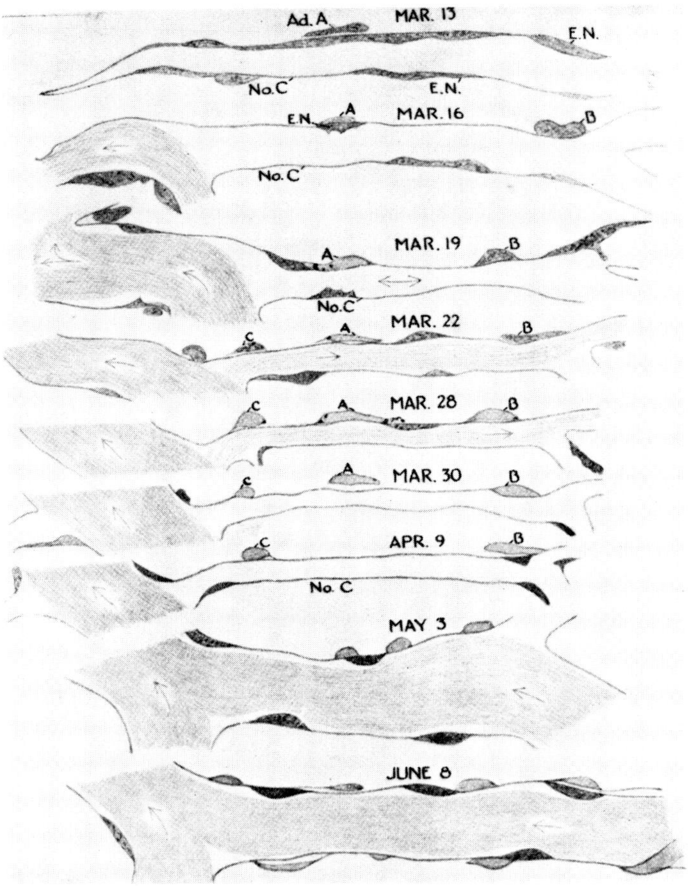

Fig. 3. Vessel growth in a rabbit ear chamber. Modified from *Clark and Clark* [7]. A capillary sprout with one adventitial cell (A) gradually develops into a venule. A, B and C mark the position of adventitial cells. Arrows indicate flow.

Fig. 2. Growth of capillaries in tadpole tails, after *Clark* [5]. Note capillaries with fast and slow flow, and capillaries without flow (dotted) in the upper part. The bottom part represents the same field in a larger tadpole 22 days later. Capillaries without flow retracted while capillaries with fast flow (marked by numbers) have new sprouts.

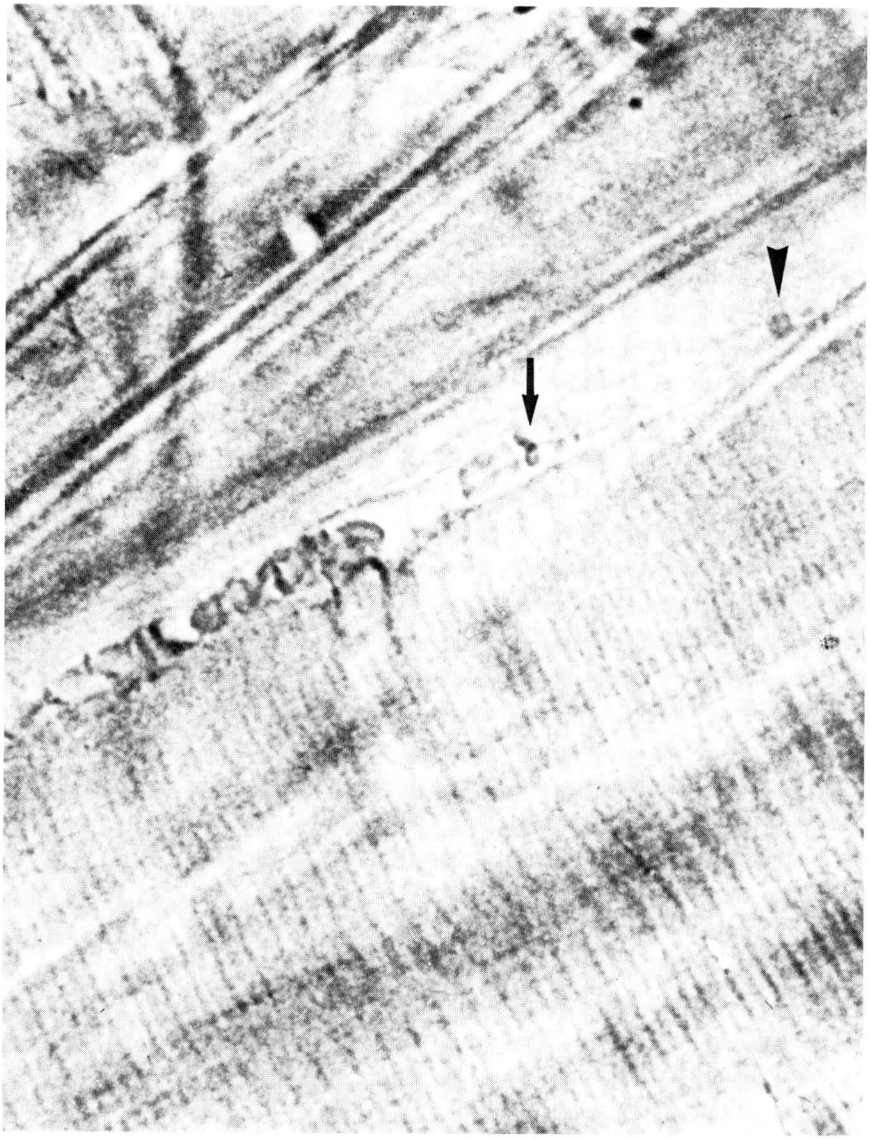

Fig. 4. Capillary sprout with two erythrocytes (↓) and one platelet (▼) in rat extensor hallucis proprius, stimulated at 10 Hz for 8 h/d for 12 days. From *Myrhage and Hudlicka* [19].

absent. *Lewis* [16] first demonstrated formation of capillary tubes from embryonic chicken endothelial cells in cultures. *Recently, Folkman and Haudenschild* [9] showed formation of capillaries from capillary endothelial cells derived from human foreskin in tissue cultures without admixture of any other cell type, but in the presence of tumour angiogenic factors. However, mechanical factors are important for growth of endothelial cells in cultures: *Folkman and Greenspan* [8] noticed that flat cells in cultures are more sensitive to several growth factors than spherical cells. *Folkman and Moscona* [10] suggested that the combination of cell geometry and some mitogenic factors may be responsible for capillary proliferation in vitro, and *Gospodarowicz et al.* [12] supposed that a similar combination may be important for capillary growth in vivo. *Ryan* [21] also showed the importance of both factors, and we hopefully will learn much more about one or the other in the course of this meeting.

References

1 Billroth, T.: Untersuchungen über die Entwicklung der Blutgefäße (Berlin 1856).
2 Boerhaave, H.: Aphorismi de cognoscendis et curandis morbis (Norimbergae 1747).
3 Brånemark, P. I.: Capillary form and function. Biblthca anat., vol. 7, pp. 9–28 (Karger, Basel 1965).
4 Chapman, W. B.: The effect of heart beat upon the development of the vascular system in the chick. Am J. Anat. *23:* 175–203 (1918).
5 Clark, E. R.: Studies on the growth of blood vessels in the tail of the frog larvae. Am. J. Anat. *23:* 37–88 (1918).
6 Clark, E. R.; Clark, E. L.: Microscopic observations on the growth of blood capillaries in the living mammal. Am. J. Anat. *64:* 251–299 (1939).
7 Clark, E. R.; Clark, E. L.: Microscopic observations on the extraendothelial cells of living mammalian blood vessels. Am. J. Anat. *66:* 1–49 (1940).
8 Folkman, J.; Greenspan, H. P.: Influence of geometry on control of cell growth. Biochim. biophys. Acta *417:* 211–231 (1975).
9 Folkman, J.; Haudenschild, C.: Angiogenesis in vitro. Nature, Lond. *228:* 551–556 (1980).
10 Folkman, J.: Moscona, A.: Roll of cell shape in growth control. Nature, Lond. *273:* 345–349 (1978).
11 Fry, D. L.: Acute vascular endothelial changes associated with increased blood velocity gradients. Circulation Res. *22:* 165–197 (1968).
12 Gospodarowicz, D.; Greenburg, G.; Birdwell, C. R.: Determination of cellular shape by the extracellular matrix and its correlation with the control of cellular growth. Cancer Res. *38:* 4155–4171 (1978).
13 Hunter, J.: Treatise on the blood, inflammation and gunshot wounds (London 1794).
14 Knower, H. M. E.: Effects of early removal of the heart and arrest of the circulation on the development of frog embryos. Anat. Rec. *7:* 161–165 (1907).
15 Landerer: Quoted from *Marchand:* Der Process der Wundheilung (Enke, Stuttgart 1901).

16 Lewis, W. H.: The outgrowth of endothelium and capillaries in tissue culture. Bull. Johns Hopkins Hosp. *48:* 242–253 (1931).
17 Loeb, J.: Über die Entwicklung von Fischembryonen ohne Kreislauf. Pflügers Arch. *54:* 525–531 (1893).
18 Meyer, J.: Über die Neubildung von Blutgefäßen in plastischen Exudaten seröser Membranen und in Hautwunden. Ann. Charité (Berlin) *4:* 41–140 (1852).
19 Myrhage, R.; Hudlicka, O.: Capillary growth in chronically stimulated adult skeletal muscle as studied by intravital microscopy and histological methods in rabbits and rats. Microvasc. Res. *16:* 73–90 (1978).
20 Platner, E. A.: Einige Beobachtungen über die Entwicklung der Capillargefäße. Müller's Arch. Anat. Physiol. Wiss. Med. 525–526 (1844).
21 Ryan, T. J.; Barnhill, R.: Physical factors and angiogenesis; in Development of the Vascular System; Ciba Fdn. Symp. *100*: 90–94 (1983).
22 Sabin, F. R.: Studies on the origin of blood vessels and of red corpuscles as seen in the living blastoderm of the chick during the second day of incubation. Contr. Embryol. Carneg. Instn. *9:* 213–259 (1920).
23 Thoma, R.: Untersuchungen über die Histogenese und Histomechanik des Gefäßsystems (Enke, Stuttgart 1893).
24 Thoma, R.: Über die Histomechanik des Gefäßsystems und die Pathogenese der Angiosklerose. Virchows Arch. path. Anat. Physiol. *204:* 1–74 (1911).

Dr. Olga Hudlicka, Department of Physiology, The Medical School, University of Birmingham, Vincent Drive, Birmingham B15 2TJ (England)

Prog. appl. Microcirc., vol. 4, pp. 9–15 (Karger, Basel 1984)

Capillary Growth in the Heart and Skeletal Muscle during Dipyridamole-Treatment and Exercise

A. Ljungqvist, G. Tornling, G. Unge, E.-B. Jurdahl, B. Larsson

Departments of Pathology and Thoracic Medicine, Karolinska Hospital, Stockholm, Sweden

Almost fifty years ago *Petrén et al.* [10] showed that physical exercise in the laboratory animal will lead to an increased capillarization of the heart and skeletal muscle. During the last decade the interest of the capillary reaction in the heart and skeletal muscle to exercise has been renewed [3–5, 11, 12] and conflicting results have been obtained, probably due to the use of various types of exercise and various methods of investigation. Most investigators have found, however, a capillary neoformation in the heart and skeletal muscle during exercise although this increased capillarization was not always sufficient to compensate for the hypertrophy of the muscle fibers involved.

In a series of investigations we have found a proliferation of capillary wall cells and further morphological evidence of a capillary neoformation in the heart of swimming exercised rats [2, 7–9, 18]. In more recent investigations the effect of exercise on the capillary network was studied also in the skeletal muscle as was the effect of dipyridamole on the capillary network of both the heart and skeletal muscle [1, 14–17]. These investigations are reviewed in the present report.

Material and Methods

Male and female Sprague-Dawley rats of different ages were used. The animals were given a standard laboratory diet containing 0.4 % sodium chloride and tap water ad libitum throughout the experimental periods. The experimental animals were divided into four main groups according to their treatment and the investigative measures.

Group I: These animals were swimming exercised for three weeks followed by investigation of the vascular reaction in the skeletal muscle. For this the DNA synthesis was measured, the DNA-synthesizing cells identified and the degree of capillarization morphometrically determined as described below.

Group II: These animals were given dipyridamole for three weeks after which they were examined in the same way as the rats in group I.

Group III: These animals were given dipyridamole for three weeks followed by identification of DNA-synthesizing cells in the myocardium.

Group IV: These animals were subjected to dipyridamole-treatment for six weeks followed by morphometric measurements of the capillarization in the myocardium.

Swimming exercise was performed by making the rats swim 1 hour/day, 5 days/week during three or six weeks. Water temperature was steadily kept at 37° C and the continuous swimming of the rats was secured. The animals given dipyridamole received intraperitoneal injections twice a day, 5 days/week at dosages of 0.3 mg/100 g body weight.

For each experimental group, a group of control animals was used. The rats used as controls for the swimming exercised rats were untreated whereas the rats used as controls for the dipyridamole-treated rats were given intraperitoneal injections of saline.

All animals were given a single intravenous injection of ^{3}H-thymidine (1μCi/nmole) of 100μCi/100 g body weight and killed 15 minutes later.

Identification of DNA-synthesizing cells was made by autoradiography. For this pieces of myocardial and/or skeletal muscle were taken and processed for the production of autoradiograms according to the stripping film technique. The autoradiograms were stained through the film with hematoxylin-eosin.

Measurement of DNA-synthesis was performed by determining the tissue radioactivity. For this the organ was quickly removed, weighed and frozen at –20° C. The tissue was then disintegrated in 0.5 ml $\times$ litre^{-1} PCA with a polytron ultrahomogenizer. After centrifugation the supernatant was discarded and the pellet resuspended in PCA. This washing procedure was repeated twice after which the final pellet was hydrolized twice in 0.5 ml $\times$ litre^{-1} PCA for 15 minutes and centrifuged. Samples of the clear supernatant were then used for colorimetric determination of DNA and measurement of tissue radioactivity in a liquid scintillation spectrometer (Packard).

The degree of capillarization was determined in the myocardium on electron micrographs. For this the heart was fixed by perfusion with 1.5 per cent glutaraldehyde solution buffered by Na-cacodylate to pH 7.4. After further fixation of small pieces in glutaraldehyde solution postfixation was performed in buffered OsO_4 and the pieces embedded in Epon. Electron microscopic sections were then produced, photographed and analyzed with respect to their capillaries. For this the capillary/fibre ratio was determined as it was the number of capillaries in contact with each fibre. This latter figure was further subdivided according to the number of fibres in contact with each capillary.

In the skeletal muscle the degree of capillarization was morphometrically determined on light microscope sections stained by a reticulin staining technique [13]. Photographs were taken and enlarged and 50–60 muscle fibres totally surrounded by other fibres and corresponding capillaries were analyzed from each specimen. By this method the capillary density was determined (capillaries/mm^2) as was the mean number of capillaries in contact with each fibre. The latter figure was subdivided according to the number of muscle fibres with which the capillaries were in contact.

Results

The autoradiograms for all experimental groups revealed that any labelling was restricted to capillary wall cells. Any increase in tissue

Table I. Myocardial tissue radioactivity (CPM/μg DNA) in vaiously aged awimming exercised rats after single injection of ^{3}H-thymidine

Group		CPM/μg DNA	Increase (%)
Young	control	23.75 ± 7.72	
	exper.	53.12 ± 30.29**	124
Adult	control	50.12 ± 22.54	
	exper.	74.87 ± 29.19*	49
Old	control	91.12 ± 33.05	
	exper.	100.25 ± 35.99	10

** $p < 0.01$
* $p < 0.05$

radioactivity would, therefore, reflect an increased incorporation into capillary wall cells indicative of proliferation of these cells and hence capillary neoformation [18].

Myocardium

Swimming exercise resulted in marked increase in tissue radioactivity of the myocardium in the young age group (table I). It also gave rise to an increase in the tissue radioactivity of adult rats, whereas in old rats the myocardial tissue radioactivity did not increase.

Table II. Myocardial tissue radioactivity (CPM/μg DNA) after single injection of ^{3}H-thymidine in dipyridamole-treated rats of various ages and sex

Group		Controls	Exper. animals	Increase %
Young	male	25.75 ± 7.72	40.20 ± 11.98**	69
	female	13.57 ± 6.21	16.60 ± 4.88	22
Adult	male	50.12 ± 22.54	70.40 ± 28.51	40
	female	28.50 ± 11.52	34.90 ± 11.10	22
Old	male	91.12 ± 33.05	126.25 ± 42.52	38
	female	45.42 ± 9.99	57.50 ± 20.11	26

** significantly different from control ($p < 0.01$)
x^2 (all groups) = 29.28 > x^2 ($p < 0.005$) = 28.30

Table III. Capillary/fibre ratio (C/F), mean number of capillaries per mm^2 (CD) and mean number of capillaries in contact with each fibre (CC) in the hind limb skeletal muscle in young and adult rats subjected to swimming exercise and in untrained rats

	Young		Adult	
	Untrained	Trained	Untrained	Trained
	n = 7	n = 7	n = 8	n = 8
C/F	0.98 ± 0.08	1.37 ± 0.19***	1.38 ± 0.16	1.74 ± 0.27*
CD	460 ± 64	710 ± 127***	393 ± 115	484 ± 135
CC	3.01 ± 0.24	4.03 ± 0.41***	4.08 ± 0.38	4.87 ± 0.58

* $p < 0.05$
*** $p < 0.001$

Dipyridamole treatment gave rise to a significant increase in tissue radioactivity in young male rats (table II). Such treatment also caused increase of varying degrees in young female rats, in adult male and female rats and in old male and female rats. In none of these individual groups, however, was the increase statistically significant, but if all experimental rats were grouped together and compared to the entire control group the overall increase was statistically highly significant ($p < 0.005$).

The morphometric analyses of the myocardial capillarization following swimming exercise and dipyridamole-treatment has not yet been completed.

Skeletal Muscle

Swimming exercise gave rise to a highly significant increase in capillary/fibre ratio, in capillary density and in the number of capillaries in contact with each fibre in the young age group of rats (table III). In the adult rats only the capillary/fibre ratio was significantly increased whereas the capillary density and the number of capillaries in contact with each fibre were not.

Dipyridamole treatment gave rise to a statistically significant increase in tissue radioactivity in the skeletal muscle of adult and old rats of both sexes, whereas the young age group did not react significantly (table IV). The morphometric analyses showed that the capillary/fibre ratio was increased in the dipyridamole-treated animals as was the number of capillaries in contact with each fibre and the capillary density (table V).

Table IV. Skeletal muscle tissue radioactivity in variously aged dipyridamole-treated rats after single injection of ^{3}H-thymidine

Group		Control	Exper. animals	Increase (%)
Young	male	60.62 ± 32.62	74.44 ± 19.23	22
	female	28.12 ± 14.06	45.40 ± 30.04	61
Adult	male	134.62 ± 55.19	244.00 ± 105.86**	81
	female	55.00 ± 18.82	93.80 ± 42.88*	70
Old	male	206.14 ± 99.00	376.37 ± 115.93**	82
	female	88.25 ± 25.86	181.90 ± 95.31**	106

* $p < 0.05$
** $p < 0.01$

Discussion

Earlier investigations have shown that in swimming exercise only the capillary wall cells of the myocardium will proliferate and give rise to new vascular channels [2, 14]. The proliferation of capillary wall cells, however, does not necessarily represent a formation of capillaries. It can alternatively be a feature of increasing turnover of the cells or the result of widening of existing capillaries. In the light of earlier investigations, the increased tissue radioactivity in the myocardium after ^{3}H-thymidine injection in dipyridamole-treated and swimming exercised rats is highly suggestive of the formation of a denser capillary network, but this will have to be finally proven by morphometrical analysis.

In the skeletal muscle such morphometrical analysis has been carried out and has shown that the increased tissue radioactivity after ^{3}H-thymidine injection in dipyridamole-treated and swimming exercised rats is actu-

Table V. Capillary/fibre ratio (C/F), mean number of capillaries/mm^2 (CD) and mean number of capillaries in contact with each fibre (CC) in dipyridamole-treated rats and in control rats

	Controls	Dipyridamole
C/F	0.98 ± 0.08	1.22 ± 0.08**
CD	462 ± 68	561 ± 87*
CC	3.01 ± 0.24	3.68 ± 0.20**

* $p < 0.05$
** $p < 0.01$

ally a feature of capillary proliferation with the formation of a denser capillary network.

The present investigation does not throw any light on the trigger mechanism for capillary wall cell proliferation and capillary neoformation. Increased blood flow as such has, however, been suggested as a stimulus for capillary neoformation [6]. This may well have taken place both in the exercised animals and in those treated with dipyridamole. In both these experimental models, the myocardial and skeletal muscle vessels will dilate. Any dilation of a vessel will result in an increased tangential stress of the vessel wall which could be expected to induce an endothelial proliferation resulting in capillary neoformation. This will, however, have to be further analysed in future experiments.

Summary

The reaction of the capillaries in the myocardium and skeletal muscle was studied in rats subjected to swimming exercise and in rats given dipyridamole. In both swimming exercised and dipyridamole-treated rats an increased proliferation of capillary wall cells in the myocardium and skeletal muscle was recorded suggesting a neoformation of capillaries in these tissues. In the skeletal muscle morphometric analysis confirmed a neoformation of capillaries. Such an analysis is also being carried out in the myocardium.

References

1 Adolfsson, J.; Ljungqvist, A.; Tornling, G.; Unge, G.: Capillary increase in the skeletal muscle of trained young and adult rats. J. Physiol., Lond. *310:* 529–532 (1981).

2 Carlsson, S.; Ljungqvist, A.; Tornling, G.; Unge, G.: The myocardial capillary vasculature in repeated physical exercise. Acta pathol. microbiol. scand., A, Pathol. *86:* 117–119 (1978).

3 Dawson, C. A.; Horvath, S. M.: Swimming in small laboratory animals. Med. Sci. Sports *2:* 51–78 (1970).

4 Harpur, R. P.: The rat as a model for physical fitness studies. Comp. Biochem. Physiol. *66A:* 553–574 (1980).

5 Hudlická, O.: Effect of training on macro- and microcirculatory changes in exercise. Exercise and Sport Sci. Reviews *5:* 181–230 (1977).

6 Hudlická, O.; Schroeder, W.: Factors involved in capillary growth in a normal adult skeletal muscle. Fed. Proc. *37:* 314 (1978).

7 Ljungqvist, A.; Unge, G.: The proliferative activity of the myocardial tissues in various forms of experimental cardiac hypertrophy. Acta pathol. microbiol. scand., A, Pathol. *81:* 233–240 (1973).

8 Ljungqvist, A.; Unge, G.; Carlsson, S.: The myocardial capillary vasculature in exercising with increased cardiac pressure load. Acta pathol. microbiol. scand., A, Pathol. *84:* 244–246 (1976).

9 Mandache, E.; Unge, G.; Appelgren, L. E.; Ljungqvist, A.: The proliferative activity of the heart tissues in various forms of experimental cardiac hypertrophy studied by electron microscope autoradiography. Virchows Arch. Abt. B Zellpath. *12:* 112–122 (1973).
10 Petrén, T.; Sjöstrand, T.; Sylvén, B.: Der Einfluß des Trainings auf die Häufigkeit der Capillaren in Herz- und Skeletmuskulatur. Arbeitsphysiol. *9:* 376–386 (1936).
11 Rakusan, K.: Quantitative morphology of capillaries of the heart. Meth. Achiev. exp. Pathol. *5:* 272–286 (1971).
12 Scheuer, J.; Tipton, C. M.: Cardiovascular adaptations to physical training. Ann. Rev. Physiol. *39:* 221–251 (1977).
13 Slidders, W.; Fraser, D. S.; Lendrum, A. C.: Silver impregnation of reticulin. J. Pathol. Bact. *75:* 478–481 (1958).
14 Tornling, G.; Unge, G.; Skoog, L.; Ljungqvist, A.; Carlsson, S.; Adolfsson, J.: Proliferative activity of myocardial capillary wall cells in dipyridamole-treated rats. Cardiovasc. Res. *12:* 692–695 (1978).
15 Tornling, G.; Adolfsson, J.; Unge, G.; Ljungqvist, A.: Capillary neoformation in skeletal muscle of dipyridamole-treated rats. Arzneimittel-Forsch. *30:* 791–792 (1980).
16 Tornling, G.; Unge, G.; Adolfsson, J.; Ljungqvist, A.; Carlsson, S.: Proliferative activity of capillary wall cells in skeletal muscle of rats during long-term treatment with dipyridamole. Arzneimittel-Forsch. *30:* 622–623 (1980).
17 Tornling, G.: Capillary neoformation in the heart of dipyridamole-treated rats. Acta pathol. microbiol. scand., C. Immunol. *90:* 269–271 (1982).
18 Unge, G.: Cardiac hypertrophy – An experimental study. Thesis, Stockholm 1973.

Prof. Dr. A. Ljungqvist, Department of Pathology, Karolinska Hospital,
S-10401 Stockholm (Sweden)

Prog. appl. Microcirc., vol. 4, pp. 16–27 (Karger, Basel 1984)

Capillary Growth in the Heart

O. Hudlicka, A. J. A. Wright, A. M. A. R. Ziada

Department of Physiology, University of Birmingham, Birmingham, England

Growth of capillaries in normal adult hearts has been described during exposure to high altitude hypoxia and in some cases of endurance training. *Turek et al.* [30] described higher capillary density (CD) and higher capillary/fibre (C/F) ratio in the right ventricle of rats exposed to high altitude (3,500 m) for four weeks, and *Grandtner et al.* [5] found greater capillarisation in animals born and maintained at a similar altitude. These authors did not see any increase in CD or C/F, in the left ventricle and concluded that low oxygen tension per se cannot be the stimulus for capillary growth. *Miller and Hale* [15] described higher capillary density in both ventricles in rats exposed to the altitude of 5,500 m for 8 weeks and suggested that polycythaemia connected with the exposure to high altitude could be responsible for capillary growth. They indeed obtained increased capillary density – comparable to that found in animals exposed to high altitude – in rats made polycythaemic by repeated transfusions and suggested that more frequent contact of red blood cells with capillary endothelium might damage the endothelial cells and stimulate their proliferation. Another factor possibly responsible for capillary growth in high altitude is increased blood flow described by *Kasalicky et al.* [8] which was greater in the right than in the left ventricle.

Increased coronary blood flow could also be responsible for capillary growth described in young, but not in adult rats, by *Tomanek* [26] and *Leon and Bloor* [9]. *Mandache et al.* [14] and *Ljungqvist and Unge* [11] found an increased incorporation of ^{3}H-labelled thymidine in heart capillaries in young rats exercised by swimming. On the other hand, *Hakkila* [6] saw a decrease in capillary density in hearts of trained guinea pigs, *Parizkova et al.* [16] did not observe any difference in CD in hearts of strenuously exercised adult rat and *Rakusan et al.* [19] found only slightly higher CD in hearts of flying pigeon compared to pigeons restricted in their movements. It is, however, known that athletic animals, such as a hare or wild rat, have a higher capillary density in the left ventricle than domestic animals of a similar species such as a rabbit or laboratory rat [33, 34].

These animals have also a lower heart rate with a longer diastole and consequently a longer period for which the terminal vascular bed is perfused. According to this assumption, it should be possible to induce capillary growth in adult hearts either by a long-term decrease in heart rate or by longer-term administration of drugs which increase coronary blood flow.

Methods of Investigation

Long-term decrease in heart rate was achieved in rabbits by bradycardial pacing [38]. The animal had bipolar pacing electrodes implanted in the right atrium, and ECG electrodes implanted under the skin 2–3 cm on either side of the midline of the thorax. The electrodes were connected to a light-weight portable stimulator [31] which delivered pacing impulses 1 msec long at about 3V using a frequency of about 55 % of the normal heart rate. It was possible to induce a premature depolarisation within the

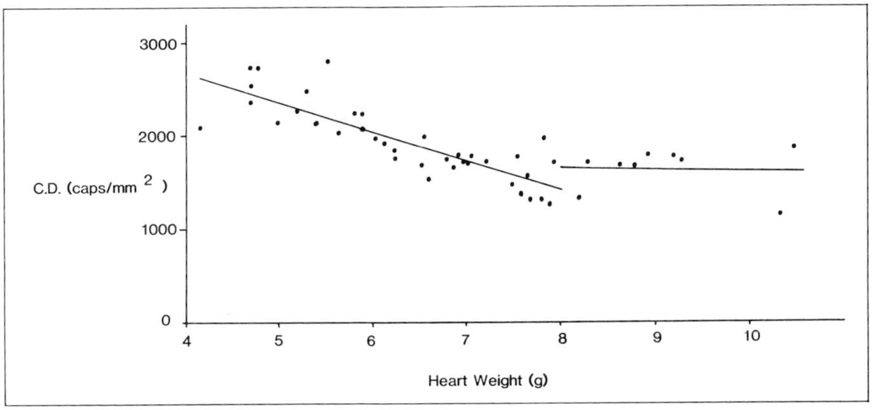

Fig. 1. Relationship between heart weight (abscissa) and capillary density (number of capillaries/mm^2, ordinate) in rabbit hearts. Data from [39].

mechanical refractory period of the heart, and thus to block every second endogenous depolarisation and associated contraction. The heart rate was thereby reduced to about the half of its resting rate. Vasodilating drugs were administered either by a long-term i. v. infusion in rabbits, for 24 h/day using portable infusion pumps [40] or in drinking water to rats. Adenosine or a xanthine derivative (HWA 285, synthetized by Hoechst, Werk Albert, Wiesbaden) were given intravenously, prazosin was given orally. The doses that would increase coronary blood flow were found in acute experiments when the drugs were infused i. v. over a period of three hours and coronary blood flow was measured with 15 μm labelled microspheres.

Capillary density was estimated in the subepicardial and subendocardial region in good cross-sections after staining for alkaline phosphatase which depicts all the capillaries present.

Changes in Capillary Density in Long-term Bradycardially Paced Hearts

Capillary density in normal hearts is related to the heart size, and it was therefore necessary to determine the relationship between the heart size and capillary density in rabbits (fig. 1). There was an inverse relationship between capillary density and the heart weight up to 8 g ($r = -0.829$, $p < 0.0001$). Heavier hearts did not show any significant correlation between the weight and capillary density ($r = -0.063$). Capillary density was evaluated from sections from both subepicardial and subendocardial layers, and since there was no obvious difference in either normal or paced hearts the results were pooled together. Pacing did not produce heart hypertrophy (the heart/body weight ratio was 0.242 in control animals and 0.241 in the paced ones), but did produce a considerable increase in capillary density (fig. 2). The number of capillaries per area was greater than in control hearts of similar size, and this was observed in all hearts paced longer than 10 days (fig. 3). Considerable increases in CD were seen within 3–4 weeks but animals paced for up to 66 days, did not show any greater increase, and a heart paced for 66 days showed only a moderately higher capillary density than predicted for a control heart of similar weight. As the size of the hearts was not changed by pacing, and in the absence of any obvious change in fibre diameter, this increase in capillary density can only be interpreted as due to growth of new capillaries.

The causes of the capillary proliferation in the bradycardially paced

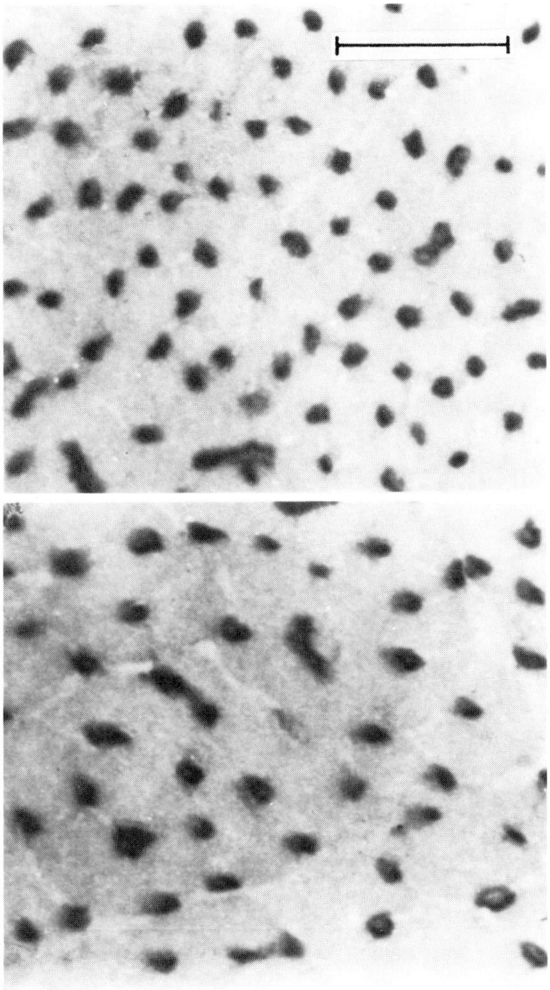

Fig. 2. Cross-section of a heart paced for 50 days (top) and from a control heart of similar size (bottom). Capillaries are depicted as black dots after staining for alkaline phosphatase. Bar = 50 microns.

hearts are not clear. *Tornling* [28] and *Ljungqvist* [10] described proliferation of capillaries in the hearts of rats exercised by swimming as well as in animals treated with dipyridamole which is known as a coronary vasodilator [35]. It is therefore possible, that a changed pattern of coronary

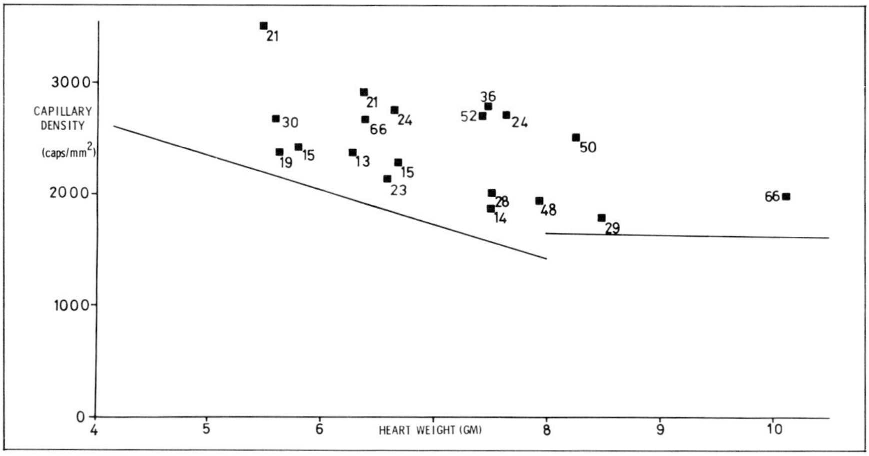

Fig. 3. Relationship between capillary density and heart weight in chronically bradycardialls-paced rabbit hearts. Regression line for normal hearts taken from fig. 1. Numbers represent days of pacing.

blood flow could be responsible for capillary growth in paced hearts. The total coronary blood flow was not greater than in control hearts [37]. However, duration of diastole was considerably prolonged. *Tillmans et al.* [25] have shown (fig. 4) that diameters of arterioles and capillaries are significantly greater in diastole than in systole, and it is possible that factors connected with wall stress and higher flow during this period might produce a slight damage to the capillary endothelium and thus stimulate its growth. If mechanical factors connected with the increased blood flow are important in the initiation of capillary proliferation, it should be possible to induce capillary growth by different drugs that increase coronary blood flow.

Changes in Capillary Density After Long-term Administration of Vasodilating Drugs

Three different drugs were used: adenosine, which is a well known coronary dilator exerting its action via interference with Ca^{++} uptake (see [2]), the xanthine derivative HWA 285 which increases coronary blood flow [7] possibly by blocking phosphodiesterase activity [23] and prazosin which is an α_1-blocker and is supposed to increase blood flow in several organs in the periphery but not in the heart [21]. The effective doses that would produce a good increase in the coronary blood pressure and heart rate were found in acute experiments when blood flow was measured with 15 μm labelled microspheres over a period of intravenous infusion of 3 hours [40]. This dose was 42 μmol/h for adenosine and 57 μmol/h for HWA 285. Both adenosine and HWA 285 increased coronary blood flow (by 38 % and 41 %, respectively) and coronary conductance (by 65 % and 33 %, respectively). Prazosin in a dose of 30 μg/h trebled blood flow in skeletal muscle, did not change coronary conductance and decreased coronary blood flow by 23 % (fig. 5). Adenosine or HWA 285 were then administered by i. v. infusion for 3–4 weeks to rabbits and a group of control rabbits was infused with saline. Prazosin was given to rats in drinking water giving consumption of about 94 μg/h (50 μg/ml, 45 ml/day) with a bioavailability of 40 % [1, 24]. Capillary density was evaluated

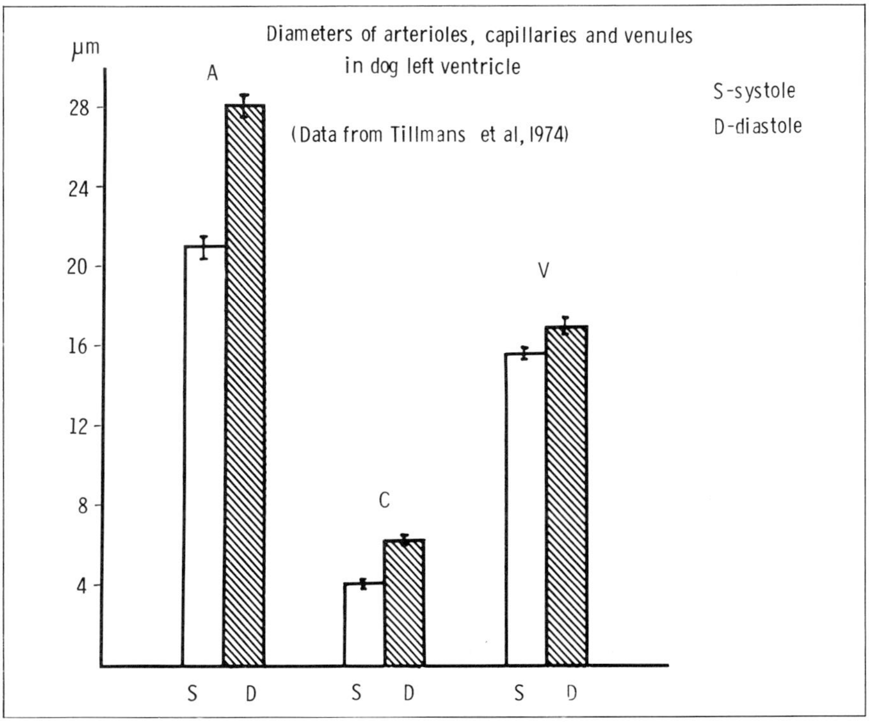

Fig. 4. Diameters of arterioles (A), capillaries (C) and venules (V) in dog hearts during systole and diastole. From data by *Tillmans et al.* [25].

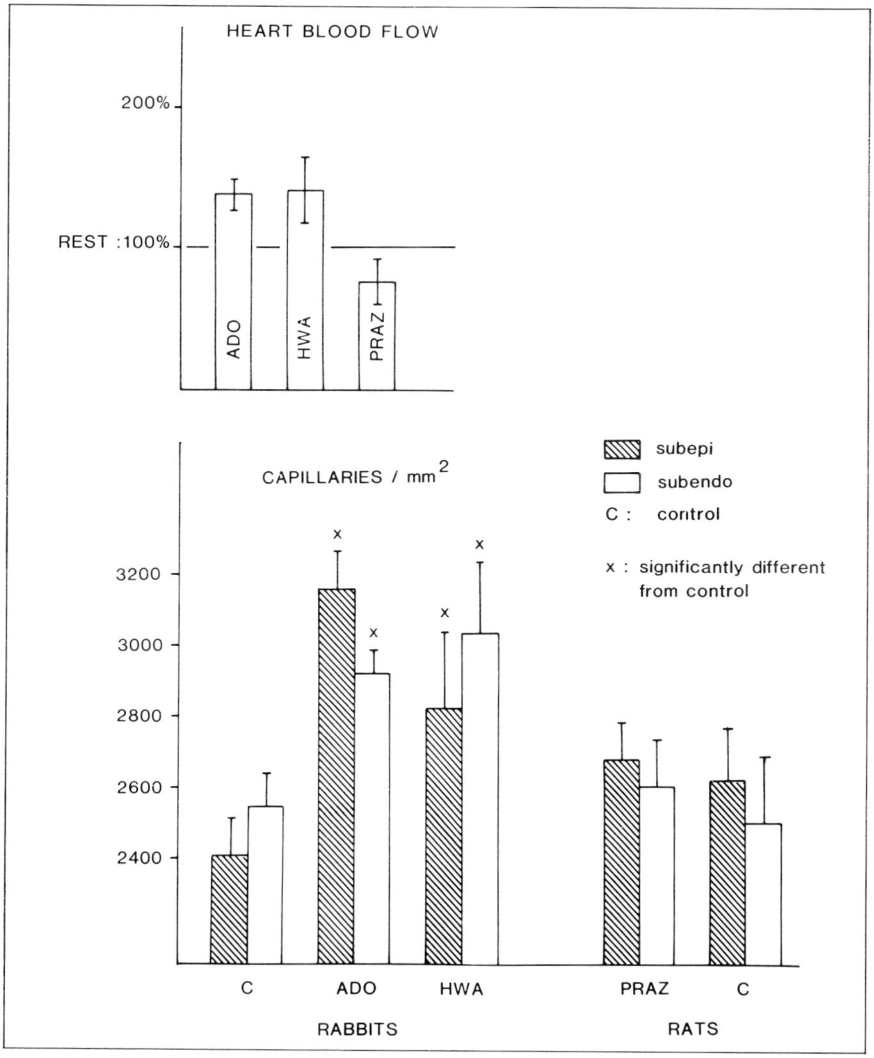

Fig. 5. Changes in coronary blood flow during 3 hours i. v. infusion of adenosine (42 µmol/h) and HWA 285 (57 µmol/h) in rabbits and prazosin (30 µg/h) in rats. Control values at rest = 100 % (left side). Capillary density in subendocardial and subepicardial regions in rabbit hearts taken from animals infused continuously by adenosine or HWA 285 for 3–5 weeks, and in rat hearts from animals given prazosin orally for 5 weeks (right side). Means ± SEM.

again in cross-sections from the subepicardial and subendocardial layer, and there was no significant difference between these two sites either in rabbit or in rat hearts. Since the heart weights in this group were very similar (5.85 ± 0.34 g in the control group, 5.40± 0.22 g in animals on adenosine and 5.77 ± 0.24 g in those infused with HWA 285) it was possible to evaluate capillary density by comparing the average data rather than the difference from the regression line between heart weight and capillary density in control animals. Both adenosine and HWA 285 produced an increase in capillary density in both the subepicardial and subendocardial layer. The increase was expressed more in the subepicardial layer with adenosine and in the subendocardial layer with HWA 285 but the variability of the data do not allow any conclusions to be drawn about preferential capillary growth in any layer with either drug. This is consistent with the finding that the distribution of flow was similar with the infusion of either adenosine or HWA 285: both drugs increased blood flow equally in both layers. These findings are consistent with the work of *Tornling et al.* [29] and *Tornling* [28] who showed capillary neoformation in hearts of rats given dipyridamole. Prazosin which did not increase coronary blood flow was without effect on capillary growth in the hearts although it increased considerably capillary density in skeletal muscles [32].

Capillary Growth in Pathological Heart Hypertrophy

Capillary growth thus can be induced in normal adult hearts by long-term bradycardial pacing or administration of drugs dilating the coronary vascular bed. The question is whether these procedures could also stimulate capillary growth under pathological conditions such as heart hypertrophy. *Mandache et al.* [13] failed to induce capillary growth in pathological heart hypertrophy by exercise which induced capillary growth in normal hearts. All types of pathological heart hypertrophy whether due to aortic constriction, valvular lesion, cold acclimatisation or hyperthyroidism show a lower capillary density [17] than control hearts. This can be expected when fibres grow considerably faster than capillaries, or when fibres grow and capillaries do not. In most cases of pathological heart hypertrophy, capillary density is lower in the subendocardial than in the subepicardial region. *Wiener et al.* [36] found it in rats with renal hypertension, *Rakusan et al.* [18] in rats with aortocaval fistula and *Gerdes et al.* [4] in hyperthyroid rats. A lower capillary density, was also found in hearts made hypertrophic by aortic valve insufficiency in rabbits [22], chronic pressure overload in cats [3] and in different types of hypertrophy in human hearts [20]. *Lund and Tomanek* [12] found lower CD in rats with heart hypertrophy induced by aortic constriction as well as in spontaneously hypertensive rats. The latter group had, however, higher CD for the same degree of

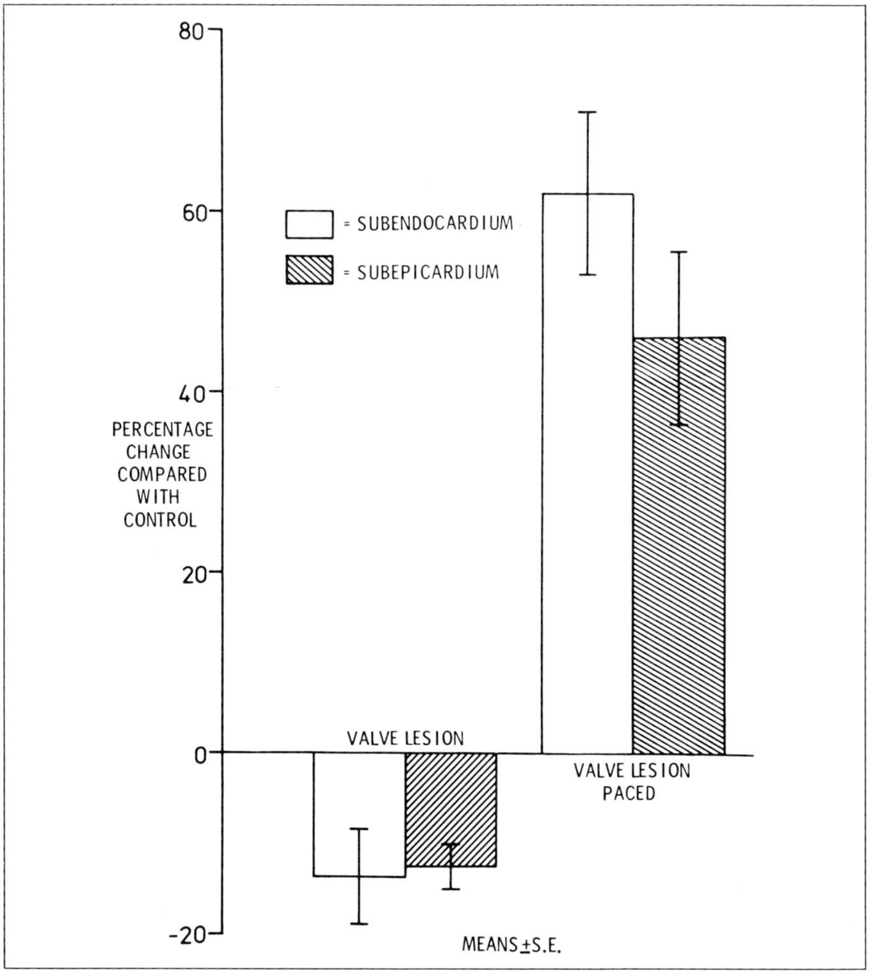

Fig. 6. Changes in capillary density in rabbit hearts made hypertrophic by aortic valve lesion and in hypertrophic and paced hearts. Zero represents the value of capillary density in control hearts of similar weight.

hypertrophy than the former indicating that some capillary growth might have occured. This was indeed confirmed by a finding of a slightly higher capillary surface area in the hearts of 15 month old SHR rats as compared to the control animals [27].

As long-term bradycardial pacing produced a greater increase in capillary density than any other procedure, it was used in rabbits whose hearts were made hypertrophic by aortic valve lesions [39]. Heart/body weight ratio was 31 ± 6 % higher than that found in control animals of similar body weight two months after aortic valve lesion, and capillary density was lower in both, the subendocardial and subepicardial region (fig. 6.). Animals in which pacing was begun one month after the lesion showed a similar increase in the heart/body weight ratio (26 ± 6 %) as the previous group, but the capillary density was considerably higher than that found in control heart of similar size, and this increase was slightly greater in the subendocardial region. This increase was also accompanied by improved stroke and minute work.

Summary

Long-term bradycardial pacing induces capillary growth in normal adult hearts, as well as in hearts made hypertrophic by aortic valve lesions. The minimum period of pacing is 10 days and the greatest increase in capillary density – about 70 % as compared with control hearts of similar size – was found in hearts paced for 3–4 weeks with no further increase in the hearts paced for a longer period. This effect might be attributed to a changed pattern in blood flow. Long-term administration of drugs (3–4 weeks) inducing coronary dilatation (adenosine or xanthine derivative) also induced capillary proliferation while prazosin was without effect on either coronary blood flow or capillary proliferation in the heart while it increased both in skeletal muscle. It is, therefore, possible that mechanical factors connected with increased blood flow stimulate capillary growth.

Acknowledgements: We gratefully acknowledge the support of the British Heart Foundation.

References

1 Bateman, D. N.; Hobbs, D. C.; Twomey, T. M.; Stevens E. A.; Rawlins, M. D.: Prazosin pharmacokinetics and concentration effect. Eur. J. clin. Pharmacol. *16:* 117–181 (1979).
2 Berne, R. M.: The role of adenosine in the regulation of coronary blood flow. Circulation Res. *47:* 807–813 (1980).
3 Breisch, E. A.; Houser, S. R.; Carey, R. A.; Spann, Y. F.; Bove, A. A.: Myocardial

blood flow and capillary density in chronic pressure overload of the feline left ventricle. Cardiovasc. Res. *14:* 469–475 (1980).
4 Gerdes, A. M.; Callas, G.; Kasten, F. H.: Differences in regional capillary distribution and myocyte sizes in normal and hypertrophic rat hearts. Am. J. Anat. *156:* 523–531 (1979).
5 Grandtner, M.; Turek, Z.; Kreuzer, F.: Cardiac hypertrophy in the first generation of rats native to simulated high altitude. Pflügers Arch. *350:* 241–248 (1974).
6 Hakkila, J.: Studies on the myocardial capillary concentration in cardiac hypertrophy due to training. Annls. Med. exp. Biol. Fenn. *33:* Suppl. 10, pp. 1–82 (1955).
7 Hudlicka, O.; Komarek, J.; Wright, A. J. A.: The effect of xanthine derivate, 1–(5' oxohexyl)–3–methyl–7–propylxanthine on heart performance and regional blood flow in dogs and rabbits. Br. J. Pharmacol. *72:* 723–730 (1981).
8 Kasalicky, J.; Ressl, J.; Urbanova, D.; Widimsky, J.; Ostadal, B.; Pelouch, V.; Vizek, M.; Prochazka, J.: Relative organ blood flow in rats exposed to intermittent high altitude hypoxia. Pflügers Arch. *368:* 111–115 (1977).
9 Leon, A. S.; Bloor, C. M.: Effects of exercise and its cessation on the heart and its blood supply. J. appl. Physiol. *24:* 485–490 (1968).
10 Ljungqvist, A.; Tornling, G.; Unge, G.; Jurdahl, E.-B.; Larsson, B.: Capillary growth in the heart and skeletal muscle during dipyridamole-treatment and exercise. Prog. app. Microcirc., vol. 4, pp. 9–15 (Karger, Basel 1984).
11 Ljungqvist, A.; Unge, G.: Capillary proliferative activity in myocardium and skeletal muscle of exercised rat. J. appl. Physiol. *43:* 306–307 (1977).
12 Lund, D. D.; Tomanek, R. J.: Myocardial morphology in spontaneously hypertensive and aortic constricted rats. Am. J. Anat. *152:* 141–147 (1978).
13 Mandache, E. G.; Unge, L.; Appelgren, E.; Ljungqvist, A.: The proliferative activity of the heart tissues in various forms of experimental cardiac hypertrophy studied by electron microscope autoradiography. Virchows Arch. Abt. B Zellpath. *12:* 112–122 (1973).
14 Mandache, E.; Unge, G.; Ljungqvist, A.: Myocardial blood capillary reaction in various forms of cardiac hypertrophy. An electron microscopical investigation in the rat. Virchows Arch. Abt. B Zellpath. *11:* 97–110 (1972).
15 Miller, A. T.; Hale, D. M.: Increased vascularity of brain, heart and skeletal muscle of polycythemic rats. Am. J. Physiol. *219:* 702–724 (1970).
16 Parizkova, J.; Wachtlova, M.; Soukupova, M.: The impact of different motor activity on body composition, density of capillaries and fibres in the heart and soleus muscles and cells migration in vitro in male rats. Int. Z. angew. Physiol. Arbeitsphysiol. *30:* 207–216 (1972).
17 Rakusan, K.: Quantitative morphology of capillaries of the heart. Meth. Achiev. exp. Pathol. vol. 5, pp. 272–286 (Karger, Basel 1971).
18 Rakusan, K.; Moravec, J.; Hatt, P. Y.: Regional capillary supply in the normal and hypertrophied rat heart. Microvasc. Res. *20:* 319–326 (1980).
19 Rakusan, K.; Ostadal, B.; Wachtlova, M.: The influence of muscular work on the capillary density in the heart and skeletal muscle of pigeon (Columbia livia dom.) Can. J. Physiol. Pharmacol. *49:* 167–170 (1971).
20 Roberts, J. T.; Wearn, J. T.: Quantitative changes in the capillary-muscle relationship in human hearts during normal growth and hypertrophy. Am. Heart J. *21:* 617–633 (1941).
21 Sheridan, D. M.; Penkoske, P. A.; Burton, E. S.; Corr, P. B.: Alpha adrenergic contributions to dysrhythmia during myocardial ischaemia and reperfusion in cats. J. clin. Invest. *65:* 161–171 (1980).
22 Shipley, R. A.; Shipley, L. J.; Wearn, J. T.: Capillary supply in normal and hypertrophied hearts of rabbits. J. exp. Med. *65:* 29–42 (1937).

23 Stefanovich, V.: Effect of 3,7-Dimethyl-1-(5oxohexyl) xanthine and 1-Hexyl-3,7-dimethyl xanthine on cyclic AMP phosphodiesterase of the human umbilical cord vessels. Res. Commun. chem. Pathol. Pharmacol. *5:* 3, 655–662 (1973).
24 Taylor, J. A.; Twomey, T. M.; Schach von Wittenau, M.: The metabolic fate of prazosin. Xenobiotica *7:* 357–364 (1977).
25 Tillmans, H.; Ikeda, S.; Hansen, H.; Sarma, J. S.; Fauvel, J. M.; Bing, R. J.: Microcirculation in the ventricle of the dog and turtle. Circulation Res. *34:* 561–569 (1974).
26 Tomanek, R. J.: Effects of age and exercise on the extent of the myocardial capillary bed. Anat. Rec. *167:* 55–62 (1970).
27 Tomanek, R. J.; Searls, J. C.; Lachenbruch, P. A.: Quantitative changes in the capillary bed during developing peak and stabilized hypertrophy in a spontaneously hypertensive rat. Circulation Res. *51:* 295–304 (1982).
28 Tornling, G.: Capillary neoformation in the heart and skeletal muscle during dipyridamole-treatment and exercise. Acta pathol. microbiol. scand., A. Pathol., Suppl *278:* 5–57 (1982).
29 Tornling, G.; Unge, G.; Skoog, L.; Ljungqvist, A.; Carlsson, S.; Adolfsson, J.: Proliferative activity of myocardial capillary wall cells in dipyridamole-treated rats. Cardiovasc. Res. *12:* 692–695 (1978).
30 Turek, Z.; Grandtner, M.; Kubat, K.; Ringnalda, B. E.; Kreuzer, F.: Arterial blood gases, muscle fibre diameters and intercapillary distance in cardiac hypertrophy of rats with an old myocardial infarction. Pflügers Arch. *376:* 209–215 (1978).
31 Tyler, K. R.; Wright, A. J. A.: Light-weight portable stimulators for stimulation of skeletal muscles at different frequencies and for cardiac pacing. J. Physiol., Lond. *307:* 6–7P (1980).
32 Tyler, K. R.; Ziada, A. M. A. R.: Factors involved in capillary redistribution in rat fast muscle following nerve crush and reinnervation. J. Physiol., Lond. *338:* 58–59P (1983).
33 Wachtlova, M.; Rakusan, K.; Poupa, O.: The coronary terminal vascular bed in the heart of the hare (Lepus europeus) and the rabbit (Oryctolagus domesticus). Physiologia bohemoslov. *14:* 328–331 (1965).
34 Wachtlova, M.; Rakusan, K.; Roth, Z.; Poupa, O.: The terminal vascular bed of the myocardium in the wild rat (Rattus norvegicus) and the laboratory rat (Rattus norvegicus lab.) Physiologia bohemoslov. *16:* 548–554 (1967).
35 Winbury, M. M.; Howe, B. B.; Weiss, H. R.: Effect of nitroglycerin and dipyridamole on epicardial and endocardial oxygen tension – further evidence for redistribution of myocardial blood flow. J. Pharmac. exp. Ther. *176:* 184–199 (1971).
36 Wiener, J.; Giacomelli, G.; Loud, A. V.; Anversa, P.: Morphometry of cardiac hypertrophy induced by experimental renal hypertension. Am. J. Cardiol. *44:* 919–929 (1979).
37 Wright, A. J. A.; Hudlicka, O.: Effect of chronic bradycardial pacing on capillary density, heart performance and coronary blood flow. Microvasc. Res. *17:* s76 (1979).
38 Wright, A. J. A.; Hudlicka, O.: Capillary growth and changes in heart performance induced by chronic bradycardial pacing in the rabbit. Circulation Res. *49:* 469–478 (1981).
39 Wright, A. J. A.; Hudlicka, O.: Capillary density and coronary blood flow in hypertrophic bradycardially-paced rabbit hearts. Microvasc. Res. *24:* 236 (1982).
40 Ziada, A. M. A. R.; Hudlicka, O.; Tyler, K. R.; Wright, A. J. A.: The effect of long-term vasodilatation on capillary growth and performance in rabbit heart and skeletal muscle. Submitted for publication to Cardiovasc. Res. (1984)

Dr. Olga Hudlicka, Department of Physiology, The Medical School, University of Birmingham, Vincent Drive, Birmingham B15 2TJ (England)

Prog. appl. Microcirc., vol. 4, pp. 28–38 (Karger, Basel 1984)

Vasculature as a Target for Tumour Therapy

J. Denekamp

Gray Laboratory of the Cancer Research Campaign, Northwood, England

It has long been recognized that the vasculature influences the response of tumours to therapy aimed at eradicating them. Tumours induce new vessels by producing tumour angiogenesis factor (TAF), but the new vascular bed that is evoked is insufficient to supply all the tumour cells with an adequate nutritional supply. For this reason some cells cease to progress around the cell cycle [14, 16, 19, 27]. Tumour cells enter the non-growth fraction and are then resistant to chemotherapeutic drugs which interfere with DNA synthesis or mitosis. Similarly, cells which become short of oxygen become resistant to radiation therapy: Hypoxia confers a 2.5–3.0 fold greater radioresistance, because oxygen is involved in the chemical fixation of radiation induced radicals [1, 12]. Thus the poor vascular network in tumours is believed to confer upon them a resistance to both chemotherapy and radiotherapy.

Conversely, the vascular supply in normal tissues has been implicated as one of the common pathways by which radiation damage in slow turn-over tissues may be mediated [8, 20, 23]. It has been proposed that vascular damage can lead to an avalanche of normal cell death, as a result of ischaemia when endothelial cells die and vessel occlusion occurs. Thus, the normal tissue vasculature is also believed to be an important determinant of the unwelcome side-effects of radiotherapy.

Strangely, the possibility of enhancing *tumour* damage by such an ischaemic mechanism has not been proposed as a realistic method of treating cancer until relatively recently. It now appears that hyperthermia (i.e. elevations of temperature by 5–8° C for 30–60 minutes) is more effective in damaging tumours than normal tissues because there is a specific sensitivity of tumour blood vessels, which can collapse after a mild heat treatment (for review [24]). Similarly, interventive radiology has developed as a means of cancer therapy. This involves the introduction of embolic materials via a catheter into the main supplying artery to a large tumour [31, 32]. Many different substances are used as the embolic focus, but this approach depends upon the ability to visualize and catheterize a critical arterial supply; hence tumours in kidney, ovary or liver are especially suitable for such treatment. A third vascular approach has been that of *Folkman et al.* [9, 11]. They have identified tumour angiogenesis factors (TAF) which are produced by tumour cells, and have sought methods of antagonizing these factors and thereby preventing the further production of new blood vessels [10].

More recently, a large difference between the proliferation kinetics of endothelial cells in normal and tumour blood vessels has been identified (factor of at least 40), and it has been postulated that some means can be devised whereby cytotoxic therapy can be directed at the vasculature, with the intention of producing an avalanche of tumour cell deaths [4, 6].

Autoradiographic Data

The quantitation of cell proliferation kinetics became possible in the 1950's when high resolution autoradiography, the recognition of the discrete phases of G_1, S, G_2 and M in the cell cycle, and the manufacture of a labelled specific precursor of DNA (tritiated thymidine) all occurred [18, 21, 29]. These developments made it possible to identify cells which were synthesizing DNA in preparation for the next mitosis, by simply providing a flash label of tritiated thymidine (^{3}HTdR), and studying which cells had incorporated the tracer by autoradiography of thin tissue sections. This technique has been widely used to study the proliferation kinetics of tumour cells in both experimental tumours and in human tumours (see [25] for review). However, until recently, little attention was paid to the stromal elements which were necessary to support the neoplasm. *Tannock* [28] studied ^{3}HTdR uptake in both tumour and endothelial cells and deduced that vascular growth might be a factor determining the overall growth rate of experimental mammary carcinomas. In 1981 *Gunduz* [13] published a very extensive study of artificial lung metastases from another mammary tumour. He found a very similar high endothelial labelling index ($\sim$ 14%) in all tumours, regardless of the tumour size or doubling time at the time of study. *Hirst et al.* [17] studied 3 different tumours and

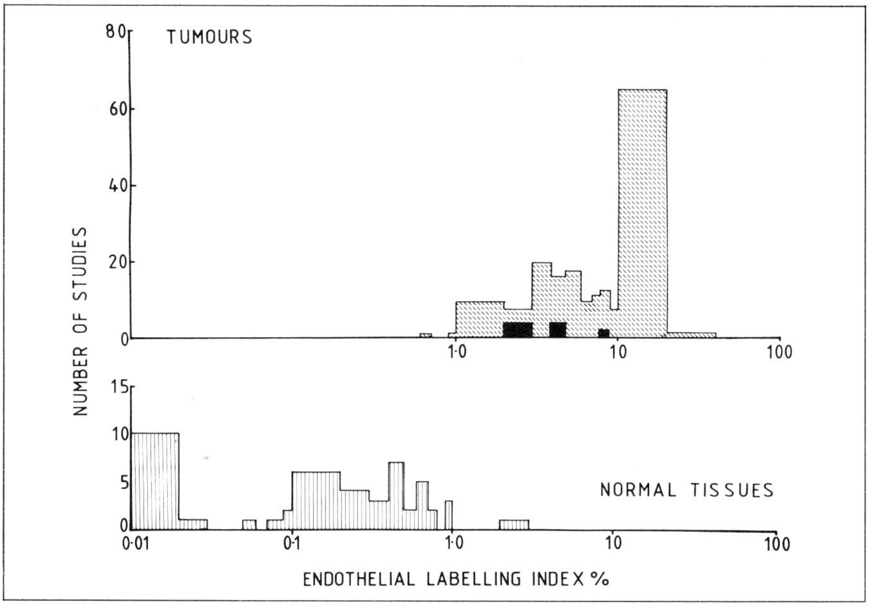

Fig. 1. Histograms showing the uptake of tritiated thymidine into endothelial cells lining the blood vessels of experimental tumours and normal tissues in rodents. There is hardly any overlap of the histograms and the median labelling indices differ by a factor of 40. The black columns represent counts of endothelium in human tumours (see text), which are similar to those in rodents.

found a range of endothelial labelling indices (4–18%) which bore little relationship to the uptake of thymidine in the adjacent tumour cells. Subsequently, a study of many different types of experimental tumour has shown that tumour endothelium is generally rapidly dividing, in marked contrast with the slowly proliferating endothelium in normal tissues [4, 6, 17]. Figure 1 illustrates this difference.

Most experimental tumours show endothelial labelling indices, after a single injection of ^{3}HTdR, in the range 10–20%. Occasionally values as low as 1–2% or as high as 32% are observed. The median value from an analysis of approximately 200 tumours is 9.0%. By contrast, the 1 hour labelling index in normal tissue endothelium is usually below 1%, often being as low as 0.01% [15]. The median value from our own and other published studies is 0.22% [6]. Thus a forty-fold difference exists in the uptake of tritiated thymidine into the vessels of tumour and normal tissue endothelium. If a similar difference could be exploited in the uptake of an

S-phase specific cytotoxic drug, this would clearly be an approach to tumour therapy worthy of further investigation. Indeed it may contribute to the effectiveness of chemotherapy on solid tumours as it is used at the present time.

Figure 1 also shows labelling index values for a variety of human tumours which have been labelled by intra-arterial infusion of ^{3}HTdR or by incubation of biopsy materials with 3HTdR in vitro. The material was provided for us by clinicians in several institutes (*C. Nervi*, Rome; *T. Hoshino*, San Francisco; *J. Kummermehr*, Munich). The labelling index values for these human tumours are similar to those found in mice and are grossly different from the normal tissue values.

We have recently extended the labelling studies in endothelium to include repeated injections of ^{3}HTdR every 8 hours over a period of 7 days [17]. In this way all cells entering DNA synthesis should be labelled, and the turnover time of the endothelium can be calculated as the time required to achieve 100% labelled cells. Figure 2 shows the data obtained

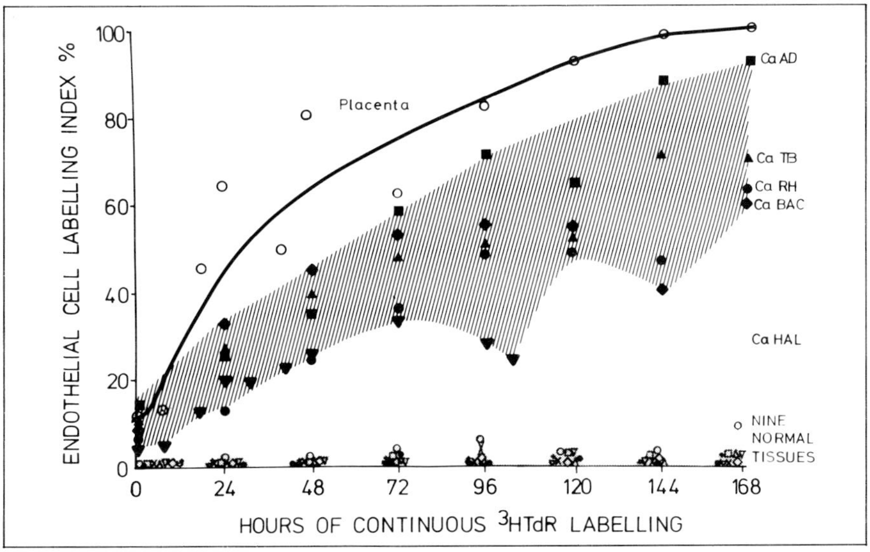

Fig. 2. Uptake of tritiated thymidine after repeated injections every 8 hours for a period of 7 days. The uptake is rapid in 5 different types of mouse tumour, and also in placenta, from 10–17 days gestation. The uptake is very slow in the nine adult normal tissues which have been studied (i. e. brain, muscle, jejunum, skin, heart, kidney, lung, liver and bladder), and is not related to the parenchymal cell turnover (Data from *Hobson and Denekamp* [17]).

for 5 experimental tumours, compared with nine normal tissues and placenta. In the tumours 35–60% of the endothelial cells are labelled within 2–3 days. In the normal tissues only 2–3% are labelled in the same time period, regardless of whether it is a rapidly dividing tissue (e.g. jejunum) or a slowly dividing tissue (e.g. heart, lung, brain). These continuous labelling data confirm that the large difference between tumours and normal tissues is real, and is not simply an artefact resulting from diurnal variations in ^{3}HTdR uptake, or from the poor statistical accuracy of 1 hour labelling indices. The uptake of ^{3}HTdR into the placenta, when

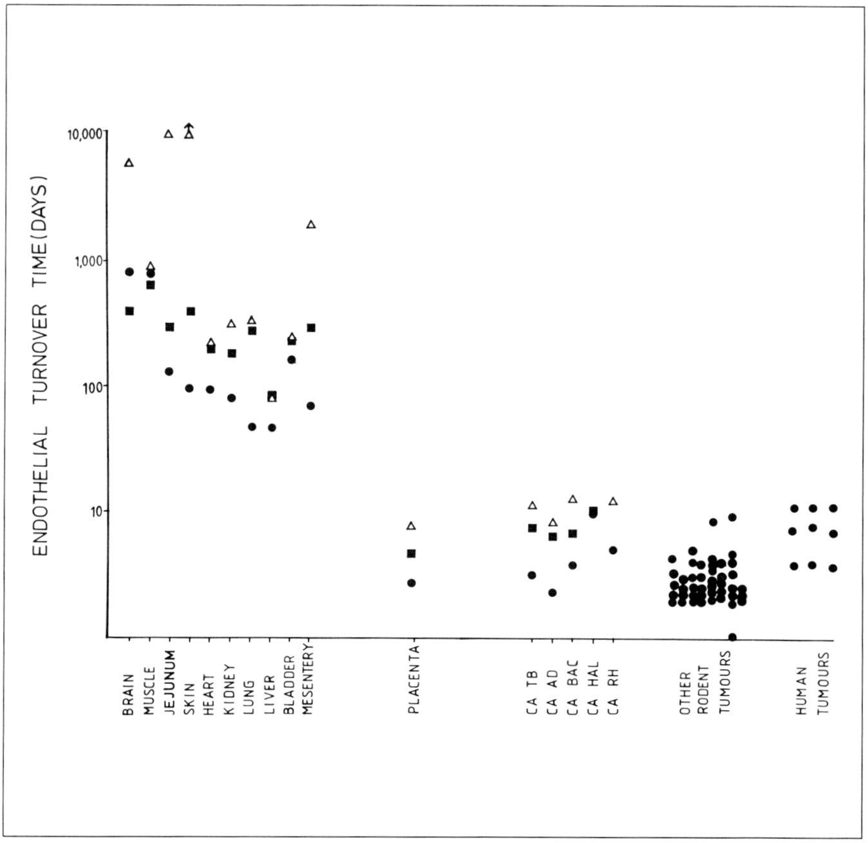

Fig. 3. Endothelial turnover times calculated from the tritiated thymidine labelling indices (L.I.). Three estimates have been made when continuous labelling data were available: ● from 1 hour L.I.; ■ from extrapolating through the 72 hour L.I.; △ from extrapolating through the 168 hour L.I. (from *Hobson and Denekamp* [17]).

labelling starts on the tenth day of gestation, is even greater than that in any of the experimental tumours. This indicates that vascular endothelium is capable of even more rapid proliferation than that which is elicited by tumour angiogenesis factors.

The 1 hour labelling indices and the continuous labelling data can both be used to estimate the proliferation rate of the endothelial cells in tumour and normal tissue vessels. Figure 3 shows turnover times (for normal vessels) or potential doubling times (for tumour vessels) calculated from such data. This figure includes all our own data on 8 steady state normal tissues, on 5 tumour types studied with continuous labelling and 19 tumour types studied with single injections. The normal tissue values are 20–2000 × longer than those in the tumour blood vessels. The proliferation rate in placenta is even faster than in most tumours. Compiled data are also shown for other studies of experimental tumours from the literature [13, 16, 28], and on bladder and mesentery [15, 26]. The turnover time of endothelial cells in human tumours is remarkably similar to that in mouse tumours, in spite of big differences in tumour size and doubling times in the two species.

It is clear from these data that a large and consistent difference in the proliferation kinetics of tumour and normal vasculature exists, which offers the potential for exploitation as a means of directing cytotoxic therapy at tumours whilst sparing normal vessels. Apart from the differences in proliferation rates, several other differences in the characteristics of tumour vessels exist which might make them more susceptible to

Table I. Differences between tumour and normal blood vessels

Normal vasculature	*Tumour vasculature*
– Is 3-dimensionally adequate	– Is 3-dimensionally inadequate leads to
– Is innervated	– nutritional deprivation
– Has collateral potential	– hypoxic cells
– Has well-constructed walls	– acidic cells
– Can increase blood flow in response to demand	– areas of necrosis
It is:	*Vessels have:*
– Resistant to radiation	– Poor wall structure
– Can withstand 90% depletion of endothelium (arterioles)	– No innervation
	– Fewer capillaries
	– More sinuses
Endothelial cells proliferate very slowly	– No collateral potential
	– Endothelial cells which proliferate rapidly

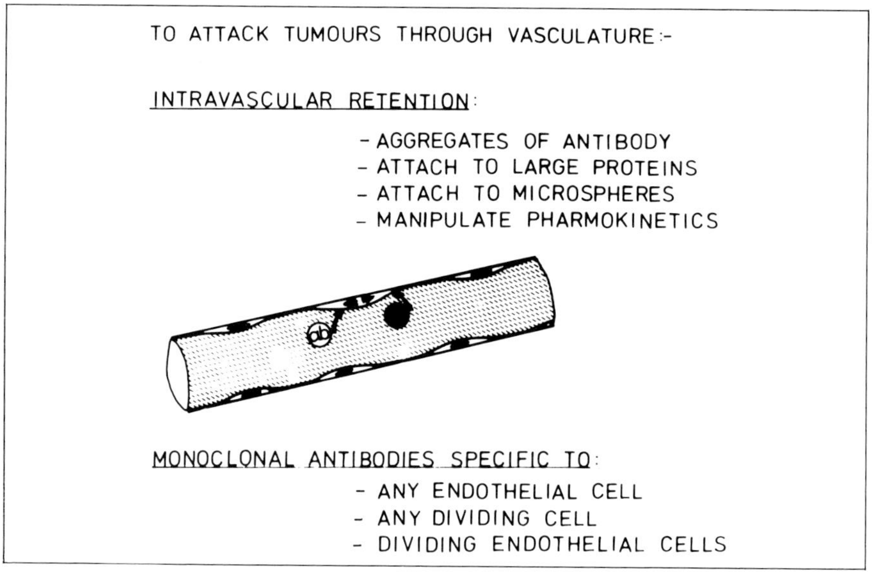

Fig. 4. Methods by which chemotherapeutic agents, radioisotopes of embolus-inducing chemicals could be targeted specifically to the proliferating endothelium of tumour blood vessels (from *Denekamp* [5]).

cytotoxic attack (table I). We have proposed that these differences might be exploited using one or more of the approaches outlined in figure 4 [5]. Monoclonal antibodies may be developed against dividing endothelial cells, and these could be used to target conjugated drugs or radioisotopes to the tumour endothelium. Alternatively, it may be possible to manipulate the pharmacokinetics of existing chemotherapeutic agents (either by means of pH, lipophilicity or by attachment to large carrier molecules) so that they are not freely diffusible throughout the body, but are retained within the vascular network. In this way the only dividing cells that would be reached by such drugs would be those immediately adjacent to the blood i.e. the endothelium, and the drug dose and side effects could be reduced.

Ischaemic Death

Death due to nutrient failure has been postulated as the cause of necrotic foci, or cords of viable tissue in seas of necrosis, at least since 1955 [30]. Measurements of the rate of production of tumour cells and the transit time across viable tumour cords allows an estimate to be made of the lifetime of naturally occurring nutrient-deprived hypoxic cells in tumours. If these cells exist as a single layer at the boundary between viable and necrotic tissue, their lifespan would only be 5–10 hours [4, 16]. We have recently performed some crude clamping experiments on mouse tumours to test the cytotoxic effectiveness of vascular occlusion [7]. Application of a clamp to the base of a subcutaneous tumour for only 2–4 hours produced significant delay in the growth of the tumour (fig. 5). For clamping times of 15 hours or longer some tumours were permanently controlled, without the addition of any other form of therapy. The necessity for several hours of occlusion is in good agreement with the 5–10 hour estimate of the natural lifespan of hypoxic cells in tumours and also with data obtained from hypoxic incubation *in vitro* [2]. These findings, together with the clinical success of interventive radiology [3, 22] encourage us to pursue the goal of aiming cytotoxic therapy at tumours indirectly, by some systemic approach to causing tumour blood vessel damage. Figure 6 illustrates the large gain in cell kill that may be achieved in this way. Each endothelial cell subtends a segment of a tumour cord containing many

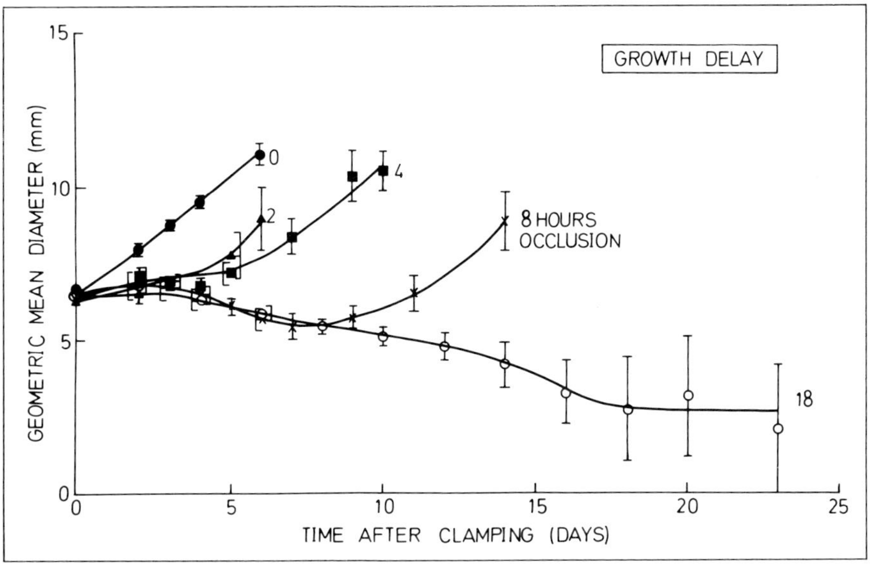

Fig. 5. Delay in tumour growth induced by the application of a clamp for 2, 4, 8 or 18 hours. A significant cell kill is achieved with 2–4 hours clamping. With occlusion for 15 hours or longer, some tumours were permanently controlled (from *Denekamp et al.* [7]).

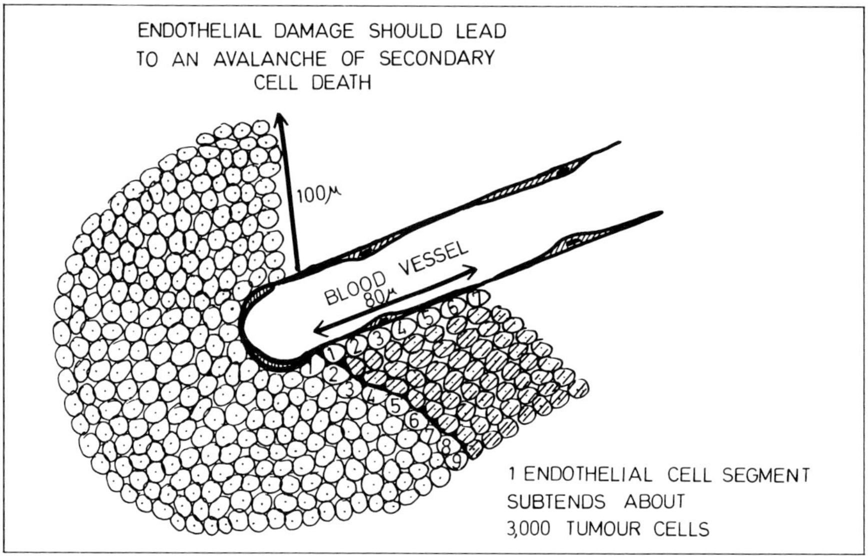

Fig. 6. Schematic illustration of how many tumour cells would be at risk if even a small segment of a capillary was occluded, so that their nutrient supply was completely lost. If an entire vessel was closed down an even greater avalanche of deaths should occur.

thousands of tumour cells. Occlusion of even small capillaries could therefore result in an avalanche of tumour cell deaths. It is anticipated that such an approach would be most effective if combined with localized radiotherapy or hyperthermia to large tumour deposits. It has the advantage, however, that it would also be effective on micrometastases, since any tumour nodule in excess of 0.15–0.25 mm diameter will be evoking a new blood supply, because nutrient diffusion will by then have become inadequate.

Conclusions

Several recent approaches to tumour therapy are based on the concept of occluding the blood supply to tumours and thereby depriving them of a nutrient supply. Any such approach is only possible if a differential can be achieved in tumour blood vessels relative to normal vessels. The large difference in proliferation parameters between such vessels has been identified as a means by which such a differential effect can be achieved.

Summary

The vascular network in tumours is inadequate, and possesses several characteristics which should make it a suitable target for anti-tumour therapy. The occlusion of individual capillaries would lead to an avalanche of cell death, amplifying the initial cell damage by a factor of at least a thousand. The rapid proliferation of endothelial cells in tumour blood vessels has been identified as the route by which systemic therapy could be targetted to the tumour.

The proliferation studies of endothelium are summarised. Uptake of tritiated thymidine after single or repeated injections has been measured using high resolution autoradiography. Tumour endothelium has a 20–2000 fold greater labelling index than endothelium in normal tissues. The only other tissue identified as having a similar rate of angiogenesis is the placenta. Nine human tumours have been studied and these fall in the same range as the rodent tumours.

The importance of this vulnerable endothelium in existing forms of cancer therapy and its potential for further exploitation are presented.

References

1 Adams, G.: Hypoxic cell sensitizers for radiotherapy; in Becker (ed.), Cancer: A Comprehensive Treatise, vol. pp. *6,* 181–224 (Plenum Press, New York 1977).

2 Born, R.; Hug, O.; Trott, K.-R.: The effect of prolonged hypoxia on growth and viability of Chinese hamster cells. Int. J. Radiat. Oncol. Biol. Phys. *1:* 687–697 (1976).

3 Chuang, V.; Wallace, S.: Current Status of transcatheter management of neoplasms. Cardiovasc. Intervent. Radiol. *3:* 256–267 (1980).

4 Denekamp, J.: Endothelial cell proliferation as a novel approach to targeting tumour therapy. Br. J. Cancer *45:* 136–139 (1982).

5 Denekamp, J.: Does physiological hypoxia matter in cancer therapy? in Steel, Adams, Peckham (eds.), The Biological Basis of Radiotherapy, pp. 139–155 (Elsevier, Amsterdam 1983).

6 Denekamp, J.; Hobson, B.: Endothelial cell proliferation in experimental tumours. Br. J. Cancer *46:* 711–720 (1982).

7 Denekamp, J.; Hill, S.; Hobson, B.: Vascular occlusion and tumour cell death. Europ. J. Cancer Clin. Oncol. *19:* 271–275 (1983).

8 Fajardo, L. F.; Stewart, J.: Capillary injury preceding radiation-induced myocardial fibrosis. Radiology *101:* 429–433 (1971).

9 Folkman, J.: Tumour Angiogenesis; in Becker (ed.), Cancer: A Comprehensive Treatise, vol. 3, pp. 355–388 (Plenum Press, New York 1975).

10 Folkman, J.; Taylor, S.; Spillberg, C.: Role of heparin in angiogenesis; in Development of the Vascular System, Ciba Fdn. Symp. *100:*132–142 (1983).

11 Folkman, J. Merler, E.; Abernathy, C.; Williams, G.: Isolation of a tumour factor responsible for angiogenesis. J. exp. Med. *133:* 275–288 (1971).

12 Fowler, J.; Denekamp, J.: A review of hypoxic cell radiosensitization in experimental tumours. Pharmacol. Ther. *7:* 413–444 (1979).

13 Gunduz, N.: Cytokinetics of tumour and endothelial cells and vascularization of lung metastases in C_3H/He mice. Cell Tiss. Kinet. *14:* 343–363 (1981).

14 Hirst, D.; Denekamp, J.: Tumour cell proliferation in relation to vasculature. Cell Tiss. Kinet. *12:* 31–42 (1979).

15 Hirst, D.; Denekamp, J.; Hobson, B.: Proliferation studies of the endothelial and smooth muscle cells of the mouse mesentery after irradiation. Cell Tiss. Kinet. *13:* 91–104 (1980).
16 Hirst, D.; Denekamp, J.; Hobson, B.: Proliferation kinetics of endothelial and tumour cells in three mouse mammary carcinomas. Cell Tiss. Kinet. *15:* 251–261 (1982).
17 Hobson, B.; Denekamp, J.: Endothelial proliferation in tumours and normal tissues: continuous labelling studies. Br. J. Cancer (in press, 1984).
18 Howard, A.; Pelc, S.: Synthesis of desoxyribonucleic acid in normal and irradiated cells and its relation to chromosome breakage. Heredity, Lond., Suppl. 6, 261–273 (1953).
19 Kligerman, M.; Heidenreich, W.; Greene, S.: Distribution of tritiated thymidine about a capillary sinusoid in a transplanted mouse tumour. Nature, Lond. *196:* 282 (1962).
20 Van der Kogel, A.: Radiation tolerance of different regions of the rat spinal cord: time-dose relationships. Radiology *122:* 505–509 (1976).
21 Pelc, S.: The stripping film technique of autoradiography. Int. J. appl. Radiat. Isotopes *1:* 1724–1728 (1956).
22 Peterson, H. I. (ed.): Tumor blood circulation: Angiogenesis, vascular morphology and blood flow of experimental and human tumors (CRC Press, Boca Raton, Florida 1979).
23 Phillips, T.: An ultrastructural study of the development of radiation injury in the lung. Radiology *87:* 49–54 (1966).
24 Song, C.: Physiological factors in hyperthermia; in Dethlefsen, Dewey (eds.): Cancer Therapy by Hyperthermia, Drugs and Radiation. NCI Monograph *61:* 169–176 (1982).
25 Steel, G.: Growth kinetics of tumours (Oxford University Press, Oxford 1977).
26 Stewart, F.; Denekamp, J.; Hirst, D.: Proliferation kinetics of the mouse bladder after irradiation. Cell. Tiss. Kinet. *13:* 75–89 (1980).
27 Tannock, I.: The relation between cell proliferation and the vascular system in a transplanted mouse mammary tumour. Br. J. Cancer *22:* 258–273 (1968).
28 Tannock, I.: Population kinetics of carcinoma cells, capillary endothelial cells and fibroblasts in a transplanted mouse mammary tumour. Cancer Res. *30:* 2470–2476 (1970).
29 Taylor, J.; Woods, P.; Hughes, W.: The organisation and duplication of chromosomes using tritium-labelled thymidine. Proc. natn. Acad. Sci. USA *43:* 122–128 (1957).
30 Thomlinson, R.; Gray, L.: The histological structure of some human lung cancers and the possible implications for radiotherapy. Br. J. Cancer *9:* 539–549 (1955).
31 Wallace, S.; Chuang, V.; Swanson, D.: Embolization of renal carcinomas; experience with 100 patients. Radiology *138:* 563–570 (1981).
32 Young, A.: Therapeutic embolization. Br. med. J. *283:* 1144–1145 (1981).

Dr. J. Denekamp, Gray Laboratory of the Cancer Research Campaign, Mount Vernon Hospital, Northwood, Middlesex HA6 2RN (England)

Prog. appl. Microcirc., vol. 4, pp. 39–53 (Karger, Basel 1984)

Angiogenesis and the Mammalian Retina

R. D. Kissun

Institute of Ophthalmology, University of London, London, England

Introduction

Angiogenesis is a fundamental biological phenomenon in many normal and pathological conditions. It is an intrinsic requirement of wound healing, in the cyclical development of the uterine mucosa, in inflammatory processes, in growing tumours and, indeed, in the developing retina. But, whereas normal growth is finely balanced to metabolic needs, the propagation of blood vessels being synchronous with tissue homeostasis, abnormal capillary growth in the normally avascular vitreous of the eye constitutes a threat to vision, as exemplified in retrolental fibroplasia (RLF) [2, 3, 49, 59], sickle cell disease [28] and diabetic retinopathy [5, 19, 30].

Most vascular lesions of the retina have been recognized for decades [34, 58] but although well documented, their aetiology remains largely unknown. There is also no satisfactory cure.

Vasculogenesis of the Retina

During intra-uterine growth, the metabolic needs of the developing retina are met initially by the choroidal circulation [63, 64] but subsequently they are augmented by the development of an inner vascular complex to meet the new demands of a rapidly differentiating retina. The development of these blood vessels has been the subject of much thought and study.

Until the early 1950's new blood vessels of the inner vascular complex were thought to have evolved by sprouting of endothelial cells from the hyaloid artery at the optic disc [45, 46]. However, observations based on tissue digest techniques showed that development occurs by an orderly migration of mesenchymal cells, liberated from the optic disc and from the adventitial tissue of the hyaloid artery and surrounding venous channels [6]. Radial migration of cells towards the ora serrata follows, and an active mitotic façade of spindle-like cells can be seen preceding a rapidly differentiating zone of organising endothelial cells, to produce the primitive vascular complex of the developing retina. Remodelling, presumably in

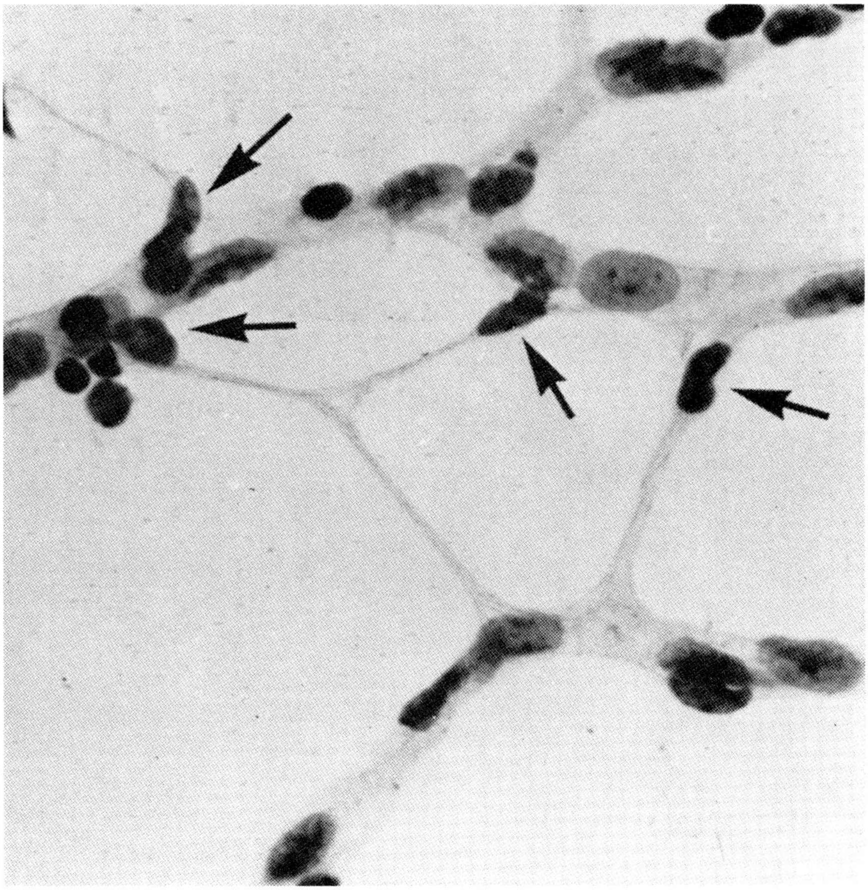

Fig. 1. Flat-mounted preparation of retinal digest from a 20-weeks-old human foetus, showing retracting capillaries and surviving cells (arrowed). Courtesy Prof. *N. Ashton.*

response to metabolic demands takes place; some connecting vessels of the polygonal mosaic retract and atrophy, whilst the residual cells may migrate to surround the newly formed endothelial capillaries where they are known as pericytes (fig. 1). There is no budding of smaller vessels at this stage, as the completed vasculature unfolds by selective atrophy and hypertrophy of the primitive mosaic. However, the deeper blood vessels of the retina develop by budding from pre-existing vessels [6]. Veins retain a large number of auxiliary vessels, that terminate at regular intervals, just short of the main arteries, creating an avascular zone which may reflect the relatively high oxygen tension on the arterial side of the retinal circulation.

It is interesting to note that earlier observations of sprouting and budding were based on injection techniques which enable visualization of established vessels, whereas digest techniques facilitate an overview of the activity of more primitive vascular cells.

Vulnerability of Retinal Blood Vessels

The blood vessels of the retina are susceptible to a range of ocular and systemic disorders that are associated with uncontrolled capillary proliferation [34, 35].

RLF and diabetic retinopathy are two well-known diseases in this category which, though having different causes, share many pathological features such as capillary malperfusion, retinal ischemia, vessel dilatation, increased vessel wall permeability, vasoproliferation into the vitreous and retinal detachment.

The Fate of Developing Blood Vessels of the Retina in High Oxygen

The identification of oxygen [36, 37] as the primary aetiological factor of RLF, led to the development of animal models in order to elucidate the pathogenesis of this disease process. Essentially, neonates from several mammalian species including mice, puppies and kittens were reared in high ambient oxygen prior to a period in their normal habitat, followed by morphogenetic studies on their retinae [2, 3, 49]. In all these instances high oxygen exposure was found to induce a proliferative retinopathy that could serve as a working model for RLF in man. It is of interest to note, that the retinal vasculature of the kitten is incomplete at term and comparable to that of the prematurely born infant.

The effect of oxygen has been elegantly described [3] and reference will be made only to a few essential points for clarification. The oxygen effect was found to be dependent upon the ambient concentration, duration of exposure and the age of the animal employed. In several experiments, kittens aged between 1 and 14 days were reared in 70–80% oxygen, over a period of 72 hours. Initially, a marked constriction of arteries and arterioles is followed by closure of the capillary bed. The main arteries and veins are the last vessels to be obliterated, to reveal an apparently avascular retina. The tunica vasculosa lentis, hyaloid artery and a few vessels at the optic disc remain unaffected (fig. 2b). A return to normal oxygen concentration is followed by revascularization of the ischemic retina, albeit in a rather disorderly fashion. These early experiments, while unravelling the aetiopathogenesis of RLF, also provided a model for studying retinal angiogenesis in other disease processes in man, such as diabetic retinopathy. More recently however, it has been suggested that the RLF model of the kitten may not truly reflect the situation in man [41]. Nevertheless, there is ample evidence that high oxygen is

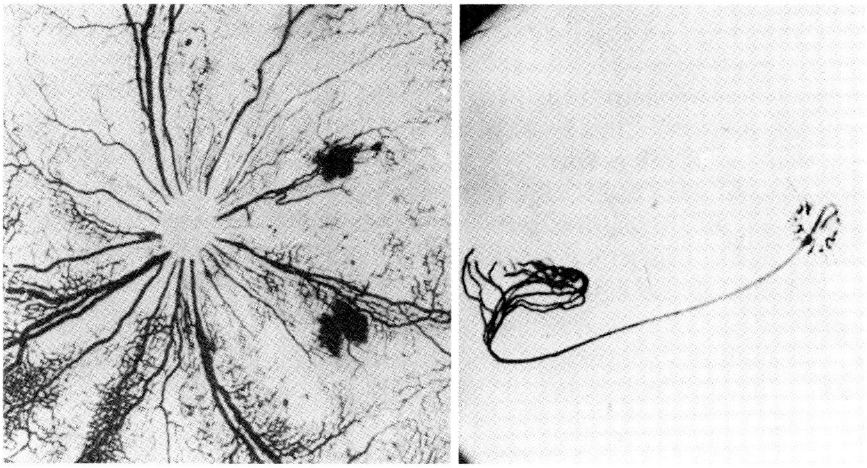

Fig. 2. a Flat-mounted preparation of a 1-day-old normal developing kitten retina. Disc region showing vascular complexes. Courtesy Prof. *N. Ashton. b* Flat-mounted preparation of a 9-days-old kitten retina following 6 days' continuous 70–80% oxygen. The entire capillary bed is obliterated apart from a few vessels around the optic disc, the hyaloid artery and tunica vasculosa lentis.

vasotoxic and produces vasoobliteration in both species. The view [51] that RLF could be a disease of prematurity instead of an oxygen effect in infants weighing less than 1 kg at birth, may be less reactionary than it seems, since it has long been known that the vulnerability to oxygen is a function of vascular tissue maturity.

Angiogenic Potential of Ocular Tissues

The observation that ischemia and other conditions that may cause a decrease in oxygen supply to the retina invariably precede proliferative retinopathy [4, 28, 57], and that destruction of the non-perfused areas of the retina by photocoagulation causes growing vessels to regress [40, 42] has suggested that a diffusible endogenous substance(s) may be formed by the ischemic retina. The relationship between restricted blood supply and blood vessel growth led to the hypothesis that the hypoxic retina produces a potent diffusible angiogenic substance(s).

In order to test this hypothesis, crude aqueous extracts of retina from normal and oxygen treated kittens following eight days survival in air, were used to determine blood vessel growth potential by introducing them into rabbit corneas [38]. Growth of blood vessels in the retina of the oxygen-treated kitten commences approximately after ten days survival in air. It is therefore assumed that the retina is rich in angiogenic substance(s). However, no significant difference could be found between the normal and oxygen treated preparations in their angiogenic capacities [38]. Subsequently other workers, using material from normal and oxygen treated puppies, reported essentially similar findings [16]. Possibly the extracts behaved as non-specific tissue culture media by providing a rich source of essential metabolites; as similar results were obtained with modified Eagles tissue culture medium [38]. However, if retinal neovascularization seen clinically is due to a tissue culture effect, it is paradoxical that only endothelial cells should be affected in an otherwise heterogeneous cell matrix. Consequently, the presence of a specific vasculogenic factor remains a reasonable proposition. The presence of blood vessel growth factors in the adult retina is probably obligatory for the maintenance of the vascular complex, as capillary growth induced by tumour angiogenic factor recedes when the factor is removed [23].

Angiogenic activity has also been demonstrated in extracts from human, bovine and feline retinas [27]; whilst fibroblast and epidermal

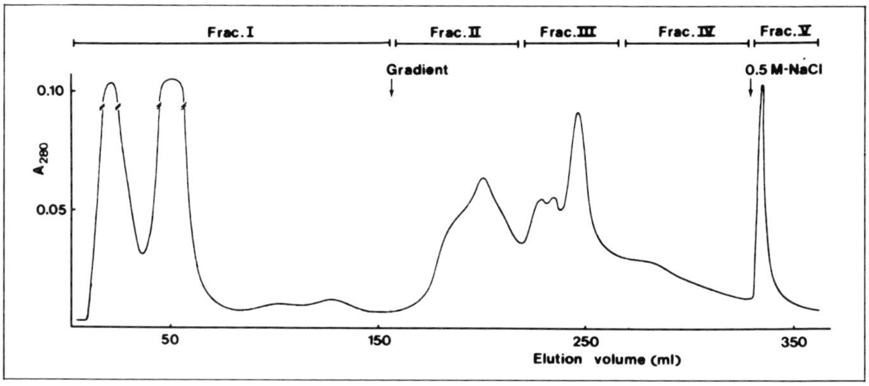

Fig. 3. DEAE cellulose chromatography of retinal homogenate.

growth factor-like properties have been described in extracts of adult bovine retina [1], vitreous, iris and choroid [9] but subsequently it was concluded that ‹Eye derived growth factor is different from other known derived growth factors› [10].

Isolation, Partial Purification and Characterization of Feline Retina Angiogenic Factor

High speed retinal supernates were purified by diethylamino-ethyl (DEAE) cellulose column chromatography, and by affinity chromatography as used for the purification of tumour angiogenic factor [60] but with slight modifications.

DEAE Cellulose Chromatography

Samples were buffered with ammonium bicarbonate, loaded on to the column and eluted with the same buffer to remove unbound material. A typical chromatogram is shown in figure 3. Fraction I contained unbound material. Bound material was eluted by buffered NaCl. Fractions II and III were eluted by a convex gradient of 0–0.3M buffered salt and fraction V by 0.5M buffered salt.

Angiogenic activity was tested on the chick chorioallantoic membrane

(CAM) assay technique. Activity was found in fractions IV and V and involves the directional growth of blood vessels towards the area of sample application (fig. 4b). No such growth is obtained in the controls (fig. 4a). Fractions I, II, and III did not give a positive response on the CAM.

Affinity Chromatography

The five fractions from DEAE chromatography were purified further by affinity chromatography. The column was packed with cyanogen bromide-activated sepharose which was bound to an antibody prepared against tumour-derived angiogenic factor. A typical affinity chromatography profile is shown in figure 5. Only fraction B gave a positive response on the CAM.

A number of aqueous based endogenous extracts from biological tissues [22, 23, 27, 29, 32, 38, 39] have been shown to give positive results on the CAM, corneal assay and in vitro stimulation of endothelial cell mitosis. The discrepancy between the reported molecular weights of the angiogenic substances (approximately 200 for rat Walker tumours and possibly for feline retina [38, 60], 2000–14 000 for wound fluid [8] and 50 000–1 000 000 for bovine retina [18]) suggests a need to establish the

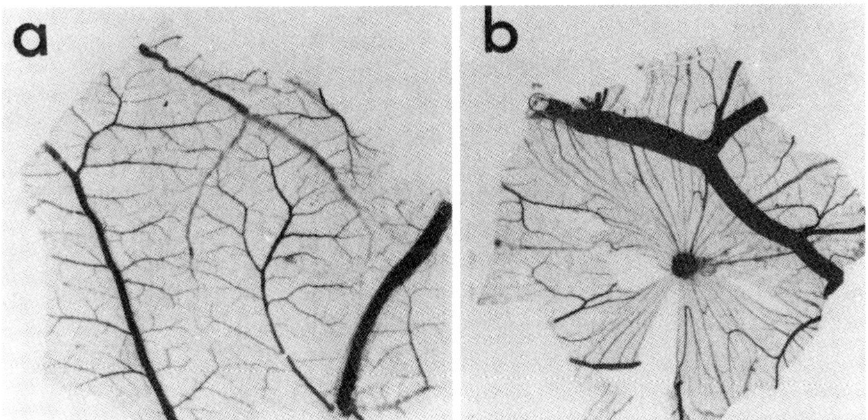

Fig. 4. Chick chorioallantoic membrane showing a control (*a*) and a test positive (*b*).

relationship between them. Whilst tumour and retinal angiogenic factors are comparable in many respects [39], the question as to whether they are identical must await further research.

Inhibitors

The absence of preretinal neovascularization in the adult retina is possibly a function of inhibitory activity in the vitreous [15, 21, 44]. In this context, it is interesting to note that both inhibitor(s) and activator(s) of angiogenesis have been found in bovine vitreous [17, 53]. Other workers have extracted non-toxic endogenous angiogenic inhibitor(s) from vitreous [14, 15, 50] and cartilage [56, 57]. These findings may prove to be a significant step towards the direct control of ocular angiogenesis.

It is reasonable to assume that hyaluronic acid, the major glycosaminoglycan in vitreous, may contain an inhibitory component, but a recent report does not support this view [44]. Nevertheless, in the chick embryo distribution of blood vessels would appear to be inversely related to hyaluronic acid concentration [20]. The inhibitory component of cartilage has been linked to antiproteolytic polypeptides [57].

Mechanisms

Lactic Acid

Ischemia and other conditions which decrease retinal blood flow are generally linked with oxygen deprivation; the presumed stimulation of anaerobic glycolysis might be expected to be coupled with increased productions and accumulation of cytoplasmic lactic acid, which has been implicated as a specific angiogenic factor. Evidence for such a process is not convincing, as no significant increase in lactic acid content has been found in diabetic rat retinas or the oxygen induced vaso-obliterated kitten retina [26]. A transient increase in lactate accumulation however, would produce intracellular acidosis which could inhibit glycolysis, decreasing not only energy supply but also activating lysomal hydrolases with possible impairment of plasma membrane and membrane permeability [62]. Nevertheless, the retina can also utilize lactate via the NADP-NADPH system [25] to provide reductive energy to drive the visual cycle (fig. 6).

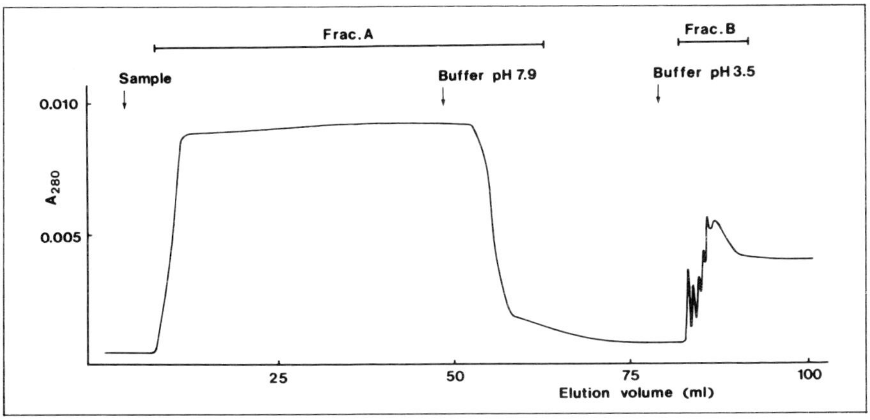

Fig. 5. Affinity chromatography of retinal extract.

Prostaglandins

Endothelial cells provide a non-thrombogenic surface and allow an unimpeded exchange of substances between the blood and tissues. This complex role in vascular haemostasis is accomplished by the production of compounds such as factor VIII-antigen [33], angiotensin converting enzyme [31], plasminogen activator and a family of prostaglandins; many

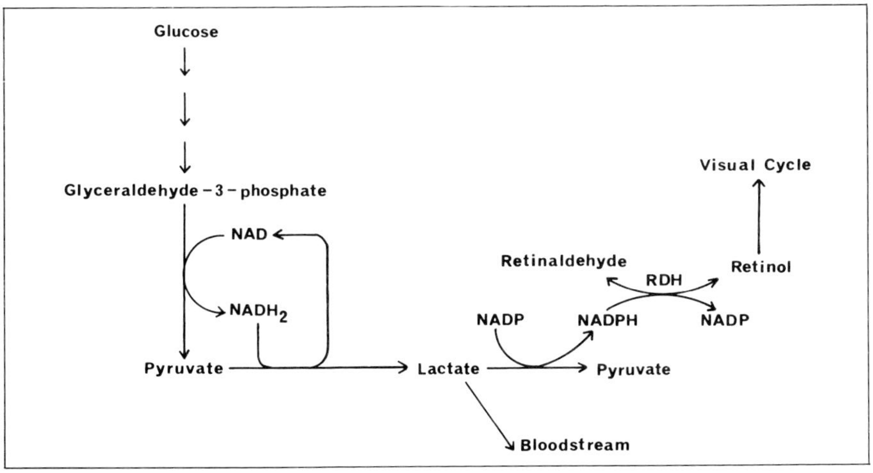

Fig. 6. Possible route for the generation of NADPH in the retina.

of which are known to have vasoactive properties [11, 43, 47, 48]. Among these, prostacyclin is a potent antithrombogenic substance and thromboxane A_2 a blood platelet aggregator. Thus modulation of the prostacyclin/thromboxane flux in favour of the latter could cause imbalance and thrombus formation. Prostaglandin E_1 has been shown to stimulate angiogenesis and copper ions seem to potentiate this effect [12, 52, 66].

Proteolytic Activity

Angiogenic substances can stimulate endothelial cells in culture to produce significantly increased amounts of two specific proteases, plasminogen activator and latent collagenase [54]. However, the activated collagenase only degrades types I, II and III collagens and not type IV, the main collagen constituent of basement membrane [35]. Endothelial cell stimulating angiogenic factor on the other hand, has been shown to activate latent collagenases against I, III and IV collagens [61]. In vitro studies have shown that bovine aortic endothelial cells in culture and their products not only stimulate blood vessel growth on the CAM, but also reproduction and movement of mesenchymal cells [29]; while other workers, using human and bovine cultured endothelial cells, have shown capillary network organisation [24]. These latter findings clearly demonstrate an inherent reversible potential for quiescent capillary endothelial cells to be reactivated. Indeed, such flexibility of a return to primitive vascular mesenchyme followed by differentiation has been postulated on an earlier occasion [5].

The behaviour of the endothelial cell during blood vessel growth has been elegantly recorded [7, 55, 65]. Focal dissolution of the basement membrane by proteolytic activity is implied. It would appear also that diffusible angiogenic stimuli induce endothelial cell transformation from a quiescent to a migratory and proliferative phase to produce new blood vessels.

Oxygen Derived Free Radicals

By analogy with experimental findings in vivo, some vascular permeability seen in inflammatory disorders may be ascribed to the formation of the oxygen derived free radical reactive oxidising species $OH^{\cdot}$, as a conse-

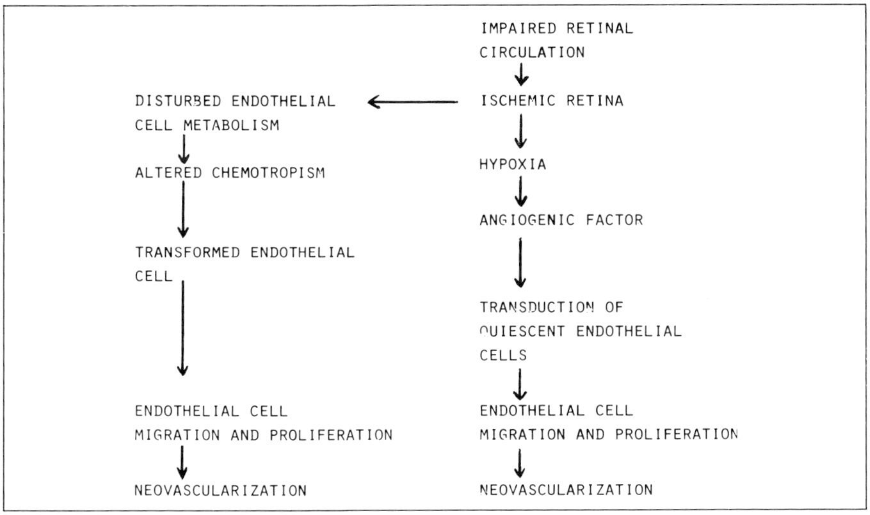

Fig. 7. Hypothetical pathways demonstrating neovascularization in the ischemic retina.

quence of the so-called ‹respiratory burst› by invading inflammatory cells [13].

Thus in some forms of retinal neovascularization, the ischemic retina triggers a cascade of events which appears to involve endothelial cell injury, lysis of the capillary wall, cell migration, proliferation and maturation of endothelial cells, with a concomitant fall in endogenous inhibitors and/or production of an angiogenic stimulus, in the quest to restore vascular haemostasis (fig. 7).

Summary

The development of new blood vessels is a common biological phenomenon. However, unregulated growth in certain disease conditions as exemplified in retrolental fibroplasia for instance, constitutes a threat to vision.

The vascular network of the mature mammalian retina appears to be differentiated from mesenchymal cells which invade the neuronal retina peripherally from the hyaloid artery. Initially, these cells join to form a polygonal mosaic of primitive blood vessels which subsequently undergo selective atrophy, surviving cells becoming integrated into the main vascular complex.

Retinal blood vessels are vulnerable to a range of ocular and systemic disorders, capillary proliferation being commonly associated with chronic inflammation and ischemia. The blood vessels of the neonate retina are unique in that they are adversely affected by high ambient oxygen. The effect is dependent upon age and is related to the concentration and duration of oxygen exposure. In the adult retina neuronal homeostasis may be affected by high oxygen, but the vascular complex appears unaffected other than a functional arteriolar constriction.

Aqueous extracts from a variety of biological tissues have been shown to stimulate endothelial cell mitosis. A low-molecular-weight angiogenic factor which has been isolated from healthy adult cat retinas and purified by diethylamino ethyl (DEAE) cellulose and affinity chromatography has a number of properties in common with a previously described tumour factor. The factor which is present in normal cat retina may be increased in hypoxic conditions.

Vitreous and cartilage extracts have been shown to inhibit endothelial cell growth.

The structural characteristics of the angiogenic factor(s) and their mechanisms of action remain to be elucidated but preliminary studies indicate that proteolytic activity may be involved. Lactic acid, free radical mechanisms and some products of arachidonic acid metabolism have also been implicated.

References

1 Arruti, C.; Courtois, Y.: Morphological changes and growth stimulation of bovine epithelial lens cells by a retinal extract in vitro. Exp. Cell Res. *117:* 283–292 (1978).

2 Ashton, N.; Ward B.; Serpell, G.: Role of oxygen in the genesis of retrolental fibroplasia: preliminary report. Br. J. Ophthal. *37:* 513–520 (1953).

3 Ashton, N.; Ward, B.; Serpell, G.: Effect of oxygen on developing retinal vessels with particular reference to the problem of retrolental fibroplasia. Br. J. Ophthal. *38:* 397–432 (1954).

4 Ashton, N.: Retinal vascularization in health and disease. Am. J. Ophthal. *44:* 7–17 (1957).

5 Ashton, N.: Oxygen and the growth and development of new retinal vessels. In vivo and in vitro studies; in Kimura, Caygil (eds.), Vascular Complications of Diabetes mellitus, pp. 3–32 (Mosby, Saint Louis 1967).

6 Ashton, N.: Retinal angiogenesis in the human embryo. Br. med. Bull. *26:* 103–106 (1970).

7 Ausprunk, D. H.; Folkman, J.: Migration and proliferation of endothelial cells in preformed and newly formed vessels during tumour angiogenesis. Microvasc. Res. *14:* 53–65 (1977).

8 Banda, M. J.; Knighton, D. R.; Hunt, T. K.; Werb, Z.: Isolation of a non-mitogenic angiogenesis factor from wound fluid. Proc. natn. Acad. Sci. USA *79:* 7773–7777 (1982).

9 Barritault, D.; Arruti, C.; Courtois, Y.: Is there an ubiquitous growth factor in the eye? Differentiation *18:* 29–42 (1981).

10 Barritault, D.; Ploüet, J.; Courtois, Y.: Purification, characterization, and biological properties of the eye derived growth factor from retina: Analogies with brain derived growth factor. J. Neurosci. Res. *8:* 477–490 (1982).

11 Bayer, B. L.; Blass, K. E.; Forster, W.: Antiaggregatory effect of prostacyclin (PGI_2) in vivo. Br. J. Pharmacol. *66:* 10–12 (1979).

12 Ben Ezra, D.: Neovasculogenic ability of prostaglandins, growth factors, and synthetic chemoattractants. Am. J. Ophthal. *86:* 455–461 (1978).

13 Björk, J.; Thaw, H. H.; Ågerup, B.; Arfors, K.-E.: Alterations in vascular permeability induced by oxygen free radicals. Acta physiol. Scand. *109:* 39A (1981).

14 Brem, S.; Brem, H.; Folkman, J.; Finkelstein, D.; Patz, A.: Prolonged tumour dormancy by prevention of neovascularization in vitreous. Cancer Res. *36:* 2807–2812 (1976).
15 Brem, S.; Preis, I.; Langer, R.; Brem, H.; Folkman, J.; Patz, A.: Inhibition of neovascularization by an extract derived from vitreous. Am. J. Ophthal. *84:* 322–328 (1977).
16 Chen, C. H.; Chen, S. C.: Angiogenic activity of vitreous and retinal extract. Investve Ophthal. vis. Sci. *19:* 596–602 (1980).
17 Chen, S. C.; Chen, C. H.: Vascular endothelial cell effectors in foetal calf retina, vitreous and serum. Investve Ophthal. vis. Sci. *23:* 340–350 (1982).
18 D'Amore, P. A.; Glaser, P. M.; Brunson, S. K.; Fenselau, A.: Angiogenic activity from bovine retina. Partial purification and characterization. Proc. natn. Acad. Sci. USA *78:* 3068–3072 (1981).
19 Dobree, J. H.: Proliferative diabetic retinopathy: evolution of the retinal lesions. Br. J. Ophthalmol. *48:* 637–649.
20 Feinberg, R. N.; Beebe, D. C.: Hyaluronate in vasculogenesis. Science *220:* 1177–1179 (1983).
21 Felton, S.; Bronn, G.; Felberg, H.; Federman, J.: Experimental ocular angiogenesis. Archs. Ophthal., Chicago *97: 1710–1713 (1979).*
22 Fenselau, A.; Watt, S.; Mello, R. J.: Tumour angiogenic factor. Purification from the Walker 256 rat tumour. J. biol. Chem. *256:* 9605–9611 (1981).
23 Folkman, J.: Tumour angiogenesis: Therapeutic implications. New Engl. J. Med. *285:* 1182–1186 (1971).
24 Folkman, J.; Haudenschild, C.: Angiogenesis in vitro. Nature, Lond. *288:* 551–556 (1980).
25 Futterman, S.; Kinoshita, J. H.: Metabolism of the retina. II. Heterogeneity and properties of lactic dehydrogenase of cattle retina. J. biol. Chem. *234:* 3173–3178 (1959).
26 Gerke, E.; Spitznas, M.; Brodde, O. E.: The role of lactic acid in neovascularization. Graefes Arch. Ophthal. *200:* 79–84 (1976).
27 Glaser, B. M.; D'Amore, P. A.; Michels, R. G.; Patz, A.; Fenselau, A.: Demonstration of vasoproliferative activity from mammalian retina. J. Cell Biol. *84:* 298–304 (1980).
28 Goldberg, M. F.: Classification and pathogenesis of proliferative sickle cell retinopathy. Am. J. Ophthal. *71:* 649–665 (1971).
29 Harris-Hooker, S. A.; Gajdusek, C. M.; Wright, T. N.; Schwartz, S. M.: Neovascular responses induced by cultured aortic endothelial cells. J. cell. Physiol. *114:* 302–310 (1983).
30 Henkind, P.: Ocular neovascularization. Am. J. Ophthal. *85:* 286–301 (1978).
31 Hial, V.; Gimbrone, M. A.; Peyton, M. P.; Wilcox, G. M.; Pisano, J. J.: Angiotensin metabolism by cultured human vascular endothelial and smooth muscle cells. Microvasc. Res. *17:* 314–329 (1979).
32 Hill, C. H.; Kissun, R. D.; Weiss, J. B.: Angiogenic factor in vitreous from diabetic retinopathy. Experientia *39:* 583–585 (1983).
33 Jaffe, E. A.; Hayer, L. W.; Nachman, R. L.: Synthesis of antihaemophilic factor antigen by cultured human endothelial cells. J. clin. Invest. *52:* 2757–2764 (1973).
34 Jampol, L. M.; Goldbaum, M. H.: Peripheral proliferative retinopathies. Surv. Ophthal. *25:* 1–14 (1980).
35 Kefalides, N. A.; Denduchis, B.: Structural components of epithelial and endothelial basement membranes. Biochemistry *8:* 4613–4621 (1969).
36 Kinsey, V. E.; Zacharias, L.: Retrolental fibroplasia. J. Am. med. Ass. *139:* 572–578 (1949).
37 Kinsey, V. E.; Jacobus, J. T.; Hemphill, F. M.: Retrolental fibroplasia: co-operative study of retrolental fibroplasia and the use of oxygen. Arch. Ophthal. *56:* 481–543 (1956).

38 Kissun, R. D.; Garner, A.: Vasoformative properties of normal and hypoxic tissue. Br. J. Ophthal. *61:* 394–398 (1977).
39 Kissun, R. D.; Hill, C. R.; Garner, A.; Phillips, P.; Kumar, S.; Weiss, J. B.: A low molecular-weight angiogenic factor in cat retina. Br. J. Ophthal. *66:* 165–169 (1982).
40 Kohner, E. M.; Shilling, J. S.; Hamilton, A. M.: The role of avascular retina in new vessel formation. Metab. Ophthalmol. *1:* 15–23 (1976).
41 Kretzer, F. L.; Hittner, H. M.; Johnson, T. A.; Mehta, R. S.; Godio, L. B.: Vitamin E and retrolental fibroplasia: Ultrastructural support of clinical efficacy. Ann. N. Y. Acad. Sci. *393:* 145–66 (1982).
42 Krill, A. E.; Archer, D.; Newell, F. W.: Photocoagulation in complications secondary to branch vein occlusion. Archs. Ophthal., Chicago *85:* 48–60 (1971).
43 Loskutoff, D. J.; Edgington, T. S.: Synthesis of a fibrinolytic activator and inhibitor by endothelial cells. Proc. natn. Acad. Sci. USA *74:* 3903–3907 (1977).
44 Lutty, G. A.; Thompson, D. C.; Gallup, G. Y.; Mello, R. J.; Patz, A.; Fenselau, A.: An inhibitor of retinal extract induced neovascularization. Investve Ophthal vis. Sci. *23:* 52–56 (1983).
45 Michaelson, I. C.: The mode of development of the vascular system of the retina with some observations on its significance for certain retinal diseases. Trans. ophthal. Soc. UK *68:* 137–180 (1948).
46 Michaelson, I. C.; Herz, N.; Lewkowitz, F.; Kertesz, D.: Effects of increased oxygen on the development of the retinal vessels: an experimental study. Br. J. Ophthal. *38:* 577–588 (1954).
47 Moncada, S.; Vane, J. R.: Pharmacology and endogenous roles of prostaglandin endoperoxides thromboxane A and prostacyclin. Pharmac. Rev. *30:* 293–331 (1979).
48 Moncada. S.; Vane, J. R.: The role of prostacyclin in vascular tissue. Fed. Proc. *38:* 66–71 (1979).
49 Patz, A.; Eastham, A.; Higginbotham, D. H.; Kleh, T.: Oxygen studies in retrolental fibroplasia. II. The production of the microscopic changes of retrolental fibroplasia in experimental animals. Am. J. Ophthal. *37:* 1511–1522 (1953).
50 Patz, A.; Brem, S.; Finkelstein, D.; Chen, C.-H.; Lutty, G.; Bennett, A.; Coughlin, W. R.; Gardner, J.: A new approach to the problem of retinal neovascularization. Ophthalmology Journal *85:* 626–637 (1978).
51 Phelps, D. L.: Vision loss due to retinopathy of prematurity. Lancet *1:* 606 (1981).
52 Raju, K. S.; Alessandri, G.; Ziche. M.; Gullino, P. M.: Caeruloplasmin, copper ions, and angiogenesis. J. natn. Cancer Inst. *69:* 1183–1188 (1982).
53 Raymond, L.; Jacobson, B.: Isolation and identification of stimulatory and inhibitory cell growth factors in bovine vitreous. Exp. Eye Res. *34:* 267–286 (1982).
54 Rifkin, D. B.; Gross, J. L.; Moscatelli, D.; Jaffe, A.: Proteases and angiogenesis: production of plasminogen activator and collagenase by endothelial cells; in Nossell, Vogel (eds.), Pathobiology of the Endothelial Cell, pp. 191–197 (Academic Press, New York 1982).
55 Schoefl, G. I.: Studies on inflammation. 111. Growing capillaries: their structure and permeability. Virchows Arch. Abt. A Path. Anat. *337:* 9–141 (1963).
56 Sorgente, N.; Kuettner, K. E.; Soble, L. W.; Eisenstein, R.: The resistance of certain tissues to invasion. 2. Evidence for extractable factors in cartilage which inhibit invasion by vascularized mesenchyme. Lab. Invest. *32:* 217–222 (1975).
57 Sorgente, N.; Kuettner, K. E.; Eisenstein, R.: The isolation, purification and partial characterization of proteinase inhibitors from bovine cartilage and aorta; in Peeters (ed.), Proceedings of the 23rd Conference on Protides of Biological Fluids, pp. 227–230 (Pergamon Press, New York 1976).
58 Stefansson, E.; Landers, M. B.; Wolbarsht, M. L.: The role of oxygen and vasodilation in diabetic retinopathy; in Friedman, L'Esperance (eds.), Diabetic Renal-Retinal Syn-

drome *2,* Prevention and Management, pp. 117–150 (Grune and Stratton, New York 1982).
59 Tery, T. L.: Extreme prematurity and fibroblastic overgrowth of persistent vascular sheath behind each crystalline lens. 1. Preliminary report. Am. Ophthal. *25:* 203–204 (1942).
60 Weiss, J. B.; Brown, R. A.; Kumar, S.; Phillips, P.: An angiogenic factor isolated from tumours: a potent low-molecular weight compound. Br. J. Cancer *40:* 493–496 (1979).
61 Weiss, J. B.: Personal communication.
62 Wildenthal, K.: Lysosomal alterations in ischaemic myocardium: result or cause of myocellular damage. J. mol. cell. Cardiol. *10:* 595–603 (1978).
63 Wise, G. N.: Retinal neovascularization. Trans. Am. Ophthal. Soc. *54:* 729–826 (1956).
64 Wise, G. N.; Dollery, C. T.; Henkind, P.: The retinal circulation (Harper, New York 1971).
65 Yamagami, I.: Electron microscopic study on the cornea. 1. The mechanism of experimental new vessel formation. Jap. J. Ophthal. *14:* 41–58 (1970).
66 Ziche, M.; Jones, J.; Gullino, P. M.: Role of prostaglandin E_1 and copper in angiogenesis. J. natn. Cancer Inst. *69:* 475–482 (1982).

Dr. R. D. Kissun, Pathology Department, Institute of Ophthalmology, 17/25 Cayton Street, London EC1V 9AT (England)

Prog. appl. Microcirc., vol. 4, pp. 54–75 (Karger, Basel 1984)

Responses of Tissue Cultured Endothelial Cells to Angiogenesis Factors – A Review

TAF and Endothelial Cells

S. Kumar, A. Keegan, A. Erroi, D. West, P. Kumar, J. Gaffney

Christie Hospital, Manchester, England

Introduction

The formation of new blood vessels is known as angiogenesis and factor(s) capable of inducing angiogenesis in vivo have been isolated from diverse sources such as tumours, retina, macrophages, lymphocytes and the synovial fluid of patients with joint diseases. We report here that tumour derived angiogenesis factor (TAF), which was angiogenic in vivo, was also mitogenic in vitro for capillary but not aortic endothelial cells. This important difference between large blood vessel and capillary derived endothelial cells was due to a lack of TAF receptors and was not the result of inactivation of TAF by aortic cells. To identify the capillary cell TAF binding site, a fluorescence activated cell sorter was used to study competitive inhibition of the binding of fluorescent labelled lectins with TAF. Quantitation of cell shape change in capillary endothelial cells suggested that the mitogenic effect of TAF might be induced by an effect on the cytoskeleton.

Preliminary data indicate that purified low molecular angiogenesis factors from tumour, retina and synovial fluid of patients with rheumatic diseases, although apparently indistinguishable from one another, might have subtle structural differences hitherto unreported.

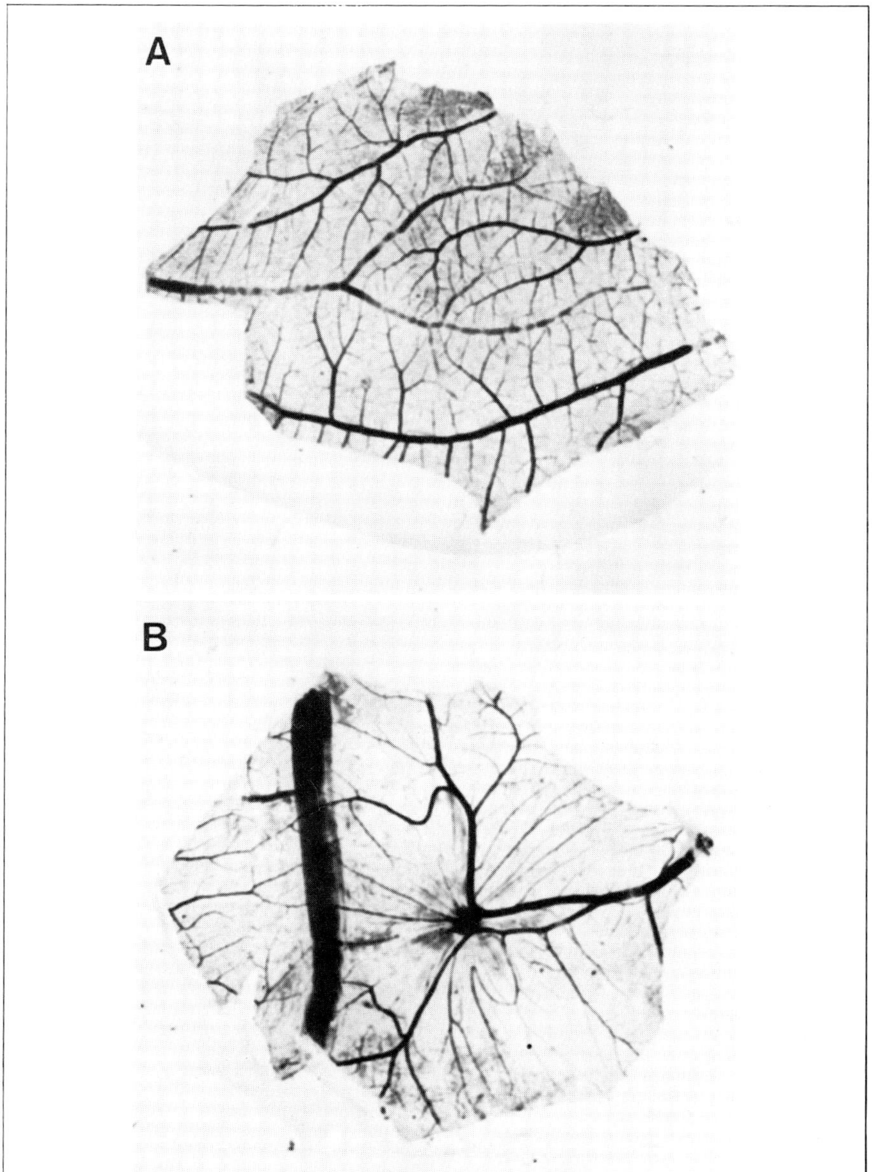

Fig. 1. A A control untreated CAM; *B* a CAM treated with tumour angiogenesis factor. The capillary density is greater than in *A* and there is a whirl of capillaries extending outwards from the centre giving it a 'spoke-wheel' appearance.

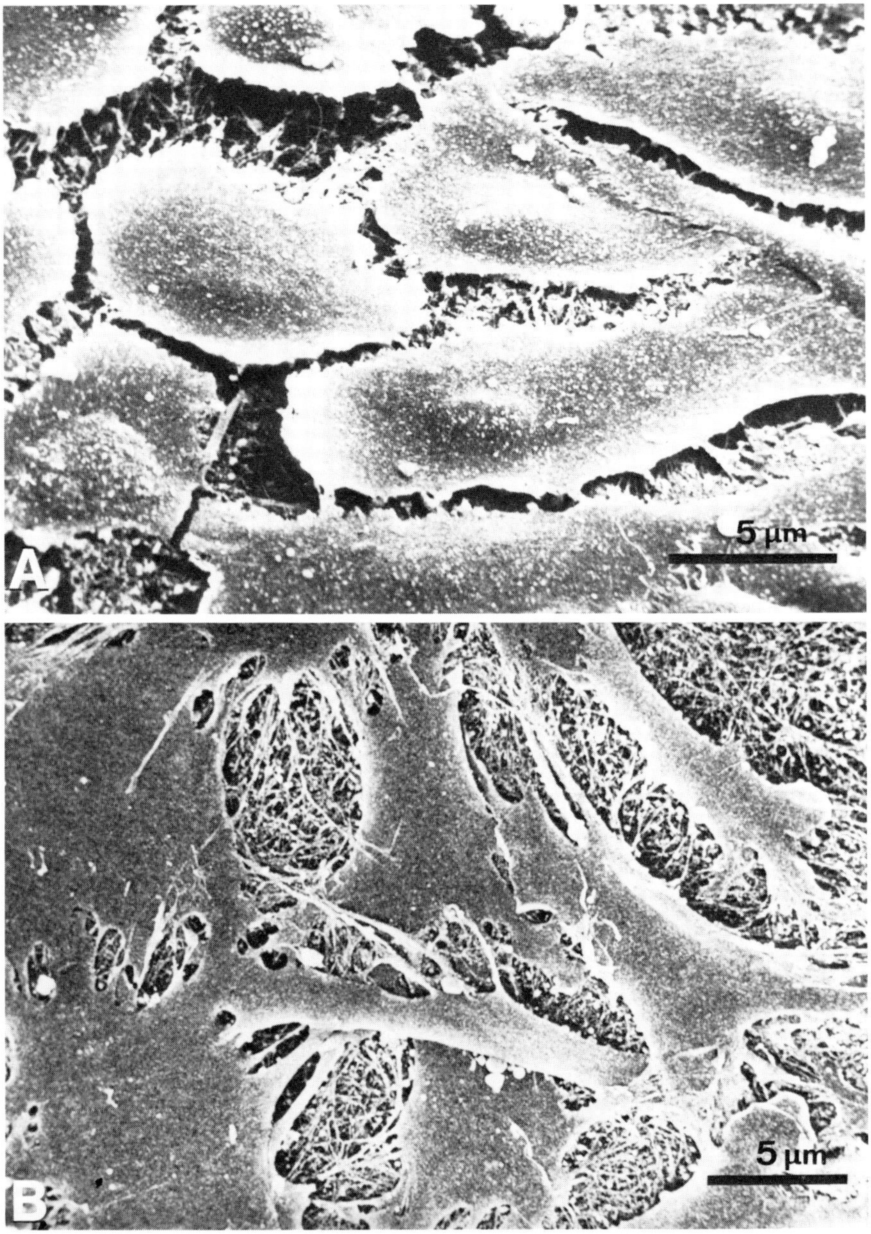

Fig. 2. Scanning EM photomicrographs of bovine capillary and aortic endothelial cells growing on collagen. *A* Aortic endothelial cells growing as a contact inhibited monolayer of

Materials and Methods

a) TAF

Extracts of Walker 256 carcinoma were purified as shown on the flow chart (fig. 11) For further details see [13]. The yield of purified material was so little that it could neither be weighed nor was it adequate for structure analysis. The purified factor is freely dialysable and its estimated molecular weight is approximately 300 daltons. To facilitate handling, it was necessary to add either lactose or albumin as a bulking agent to TAF preparations. Hence, the actual concentration of TAF in our experiments is not known, but can only be a negligible portion of the total weight. Several batches of TAF were used in the course of the present study and each was shown to induce angiogenesis on the CAM assay (fig. 1).

b) Endothelial Cells

Capillary endothelial cells (BCEC) were obtained from cow brain white matter by a method of *Phillips et al.* [14]. Bovine aortic endothelial cells (BAEC) were isolated from fresh aortas. Aortas were carefully opened, washed extensively with phosphate buffered saline and the intimal linings of the washed aortas were gently scraped with a scalpel to remove the endothelial layer. The intimal scrapings were treated with collagenase at 37° C for 15 minutes. The cell suspension was washed and inoculated into tissue culture flasks. The following day the medium was changed and thereafter every 3 days. BCEC and BAEC were used as primary cultures usually within 3 weeks of initiation.

Light microscopic examination of the cultures showed that the typical cobble-stone monolayer morphology was evident only in aortic cells. Capillary cells appeared less compact and much more flattened. Scanning EM studies (fig. 2) confirmed the light microscopic appearance viz. BCEC consistently overlapped one another and cytoplasmic processes were sometimes seen passing underneath the collagen gel. Aortic cells often

---

rectangular cells having smooth surfaces and very few short microvilli (× 5,500); *B* Cultures of capillary endothelial cells have more irregular outlines with many long processes and filopodia. Their surface is almost completely smooth (× 4,500).

had a central dome corresponding to their nucleus. BAEC cells did not overlap and exhibited fewer long cytoplasmic projections than did BCEC. Ultrastructural examination for Weibel-Palade (W-P) bodies showed that BCEC rarely contained mature W-P bodies (fig. 3) although a few pale immature W-P bodies were observed [14]. Mature W-P bodies were more common in aortic cells. Interestingly, they are more numerous in brain tumour capillaries than in normal brain capillaries [11].

Both cell types were examined for the presence of Factor VIII antigen using antibovine Factor VIII antiserum. Intense staining of BCEC and BAEC was observed (fig. 4).

γ-glutamyl transpeptidase (γ-GTP) has been shown to be present in cerebral capillaries. Our studies, using the method of DeBault [5], demonstrated significant levels of the enzyme in cryostat sections of cow brain white matter capillary endothelium used to isolate BCEC. The latter were also positive for γ-GTP (fig. 5). However, the intensity of the staining was found to decrease with age in a culture (after 3–4 weeks) and on subculturing. BAEC did not stain for γ-GTP.

High levels of both prostacyclin and angiotensin-converting enzyme were found in aortic cells but their content was negligible in BCEC (our unpublished data).

c) Culture Conditions

Proliferation: The culture conditions for both BCEC and BAEC have been described in our recent paper [10]. Briefly, subconfluent primary cultures of capillary and aortic endothelial cells were trypsinized and plated on to 35 mm^2 tissue culture dishes, coated with native collagen type I gel. 24 hrs later the medium was discarded and replaced with 2 ml of fresh growth medium with or without TAF (+ filler). Cell numbers were estimated on days 2, 3, 5 and 6 using a Coulter Particle Counter. Cultures were used in triplicate for every determination and had a standard deviation that was always less than 10 % of the mean. The significance of the differences between the means was estimated using Student's t-test.

Absorption Studies: A pilot study showed that 100 μg TAF (+ bulking agent)/dish enhanced the proliferation of BCEC.

Suspensions of capillary and aortic endothelial cells (1×10^7) were resuspended in 10 ml of tissue culture medium. Five milligrams TAF (+ filler) were added to half the cell suspension (5 ml) and the other half

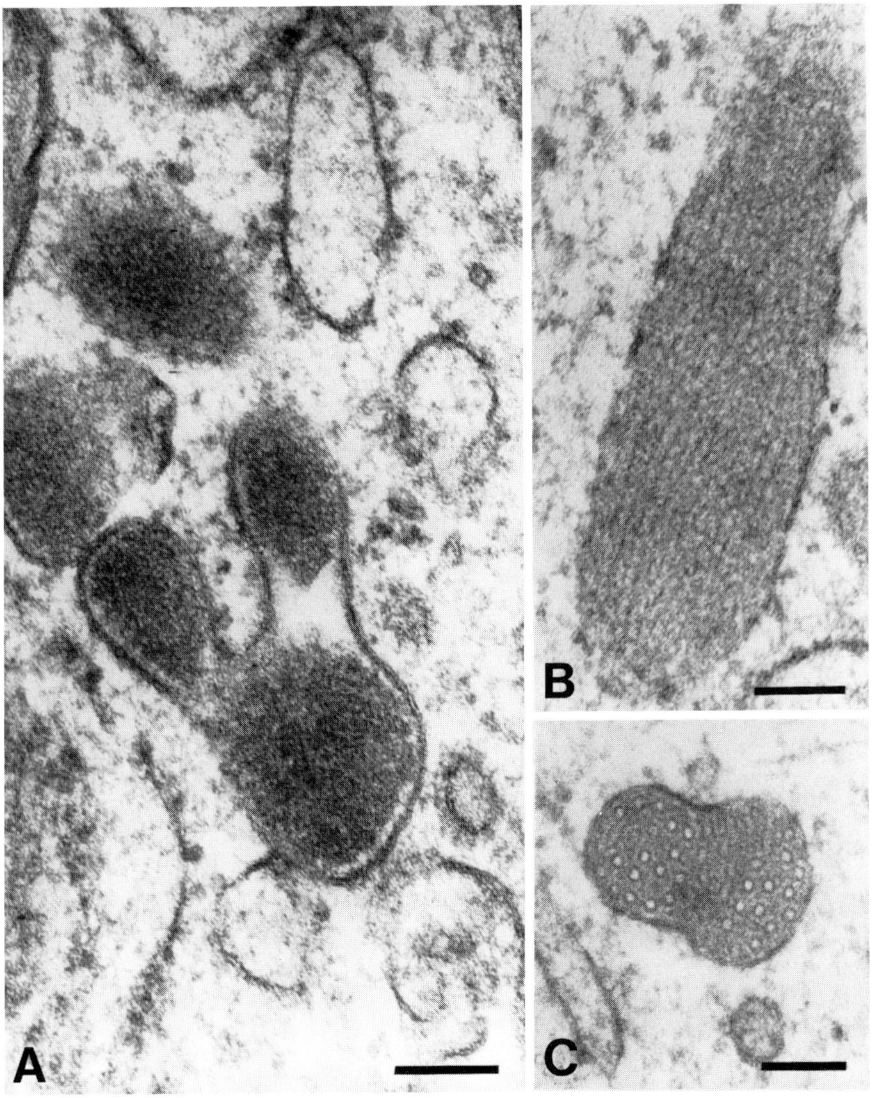

Fig. 3. Transmission EM photomicrographs of 3 Weibel-Palade bodies. *A* An unusual branching W-P body (× 186,000); *B* a longitudinal section through a typical W-P body showing its tubular structure (× 166,000); *C* a cross-section of internal tubules of a W-P body (× 150,000). Bars represent 0.1 μm.

(5 ml) acted as the control. After 4 hours incubation at 37° C, the cell suspensions were centrifuged and the cell pellets and supernatants treated as follows:

Cell Pellets: were washed in fresh medium 6 times (more than 95 % cells were viable by trypan blue exclusion) plated on to collagen gels (1.01×10^5 cells/dish) and incubated at 37° C. Cell numbers were observed over a period of 8 days.

Supernatants: were millipore-filtered and 100 μl aliquots (equivalent to 100 μg TAF (+ "filler") were added to capillary and aortic endothelial cell cultures. Cell numbers were determined over a period of 8 days.

d) Lectin Binding

Endothelial cell suspensions in serum-free medium (containing approximately 5×10^5 cells) were treated with TAF (+ filler BSA), or BSA alone, for 1 hr at 37° C. The cell suspensions were washed and fluorescein-labelled lectins (Sigma or Miles Laboratories and see legend for fig. 6) were added at the recommended concentrations. Usually 1–20 μg of lectin was added to 100 μl cell suspension in a total volume of 150 μl of PBS. In order to confirm specificity, a dose response curve with TAF and lecitin inhibitor was carried out. For instance, β-D galactose was used as an inhibitor in the ricin binding studies. Cell fluorescence was measured using a Becton and Dickinson cell sorter with help of Mr. *M. Hughes*. Usually more than 100,000 cells were analysed for each determination and the information was stored by the computer for later analysis and the production of cytograms and 3-dimensional histograms.

e) Effect of TAF Treatment on Cell Shape

Light Microscopic Study of BCEC that had attached overnight to collagen gels were exposed to different concentrations of TAF (+ filler).

Fig. 4. Localization of Factor VIII antigen by indirect immunofluorescent technique, in bovine brain capillary endothelial cells using rabbit-anti-bovine Factor VIII antiserum (a generous gift from Dr. *E. Kirby*). Low (*A*) and (*B*) high power photomicrographs showing characteristic perinuclear granular distribution of Factor VIII. A similar staining of Factor VIII in aortic endothelial cells was also seen but is not shown here.

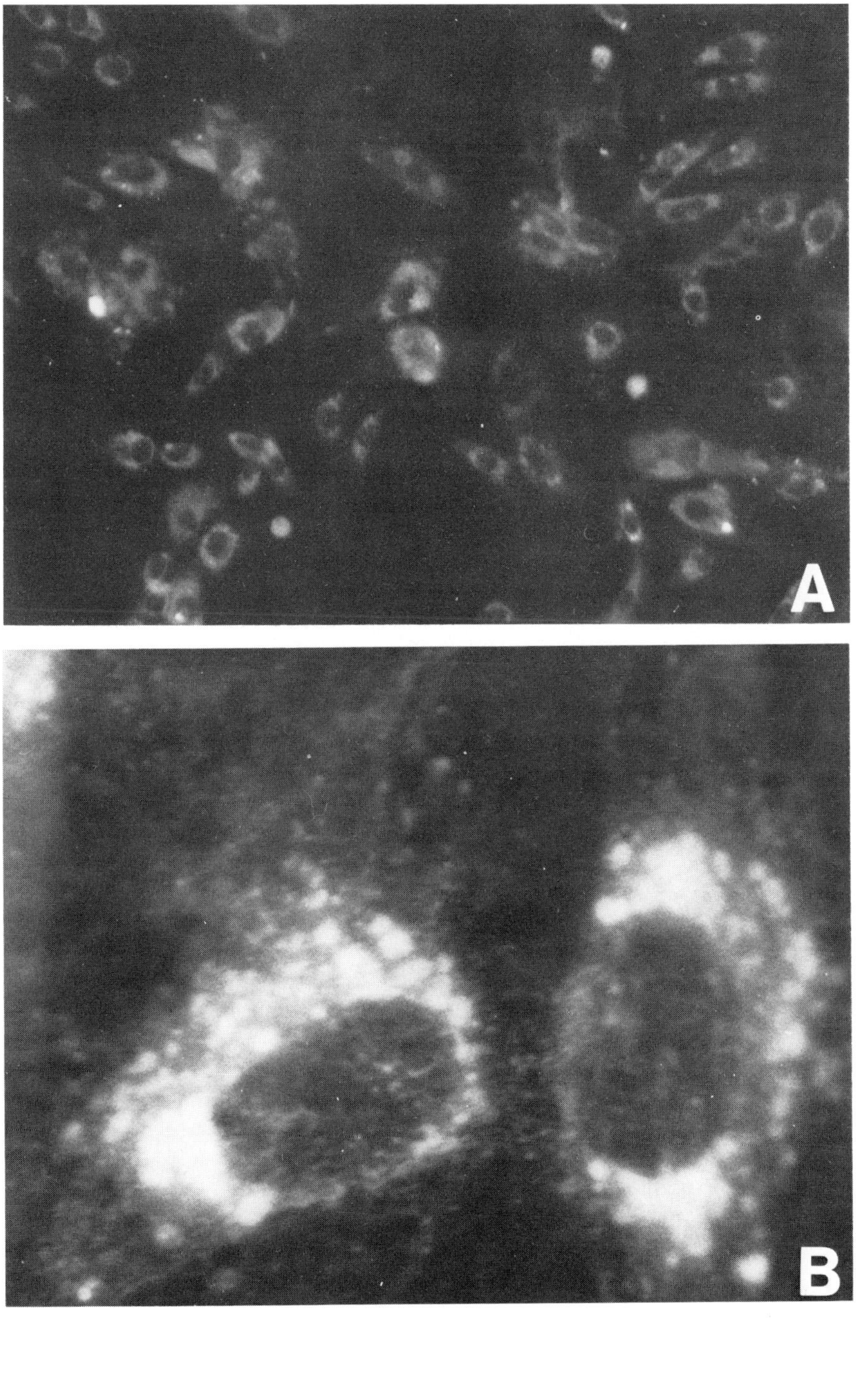
A
B

The number of fusiform and polygonal cells were counted under a Leitz inverted phase contrast microscope. Four different areas were studied from each dish and the results were analysed using a χ^2-test.

Scanning EM Study: Primary cultures of BAEC and BCEC were trypsinized and the cells were plated on collagen gels with or without TAF. Every TAF sample used was angiogenic on a CAM assay and stimulated proliferation of BCEC. Four experiments were performed with BCEC and 3 with BAEC. At various times after TAF treatment, cultures were fixed and random areas of collagen on which the cells were attached were removed with a cork borer and processed for SEM. Every cell was included for analysis and classified as: (1) rounded cells showing blebs, or microvilli on their surface, (2) semi-flattened, and (3) flattened cells (fig. 7). Several hundred cells were counted for each point of time.

The following features were quantitated and expressed as percentages of the total number of cells counted – a smooth surface, blebs and microvilli (in rounded-up cells), semi-flattened and flattened cells. The χ^2-test was used for statistical analysis.

f) Effect of TAF on Proteolytic Enzymes

Cultures of bovine capillary and aortic endothelial cells were examined for cell associated and secreted proteinase activity. ^{3}H-acetylated soluble elastin and fluorogenic substrates for elastase, trypsin, and chymotrypsin were used. Collaborative studies with Dr. *J. Reynolds* and *D. Wooley*, were also carried out to examine the production of casease and collagenase activities (details will be published elsewhere).

Results

All batches of TAF used in the present study induced angiogenesis in the in vivo CAM assay and enhanced proliferation of capillary endothelial cells only when the cells were cultured on native collagen gel. No concentration of TAF produced an increase in the number of capillary cells on

Fig. 5. γ-glutamyl transpeptidase was present in capillaries of cryostat section of bovine brain white matter *(A)* and *(B)* in tissue cultured endothelial cells (10-day-old culture).

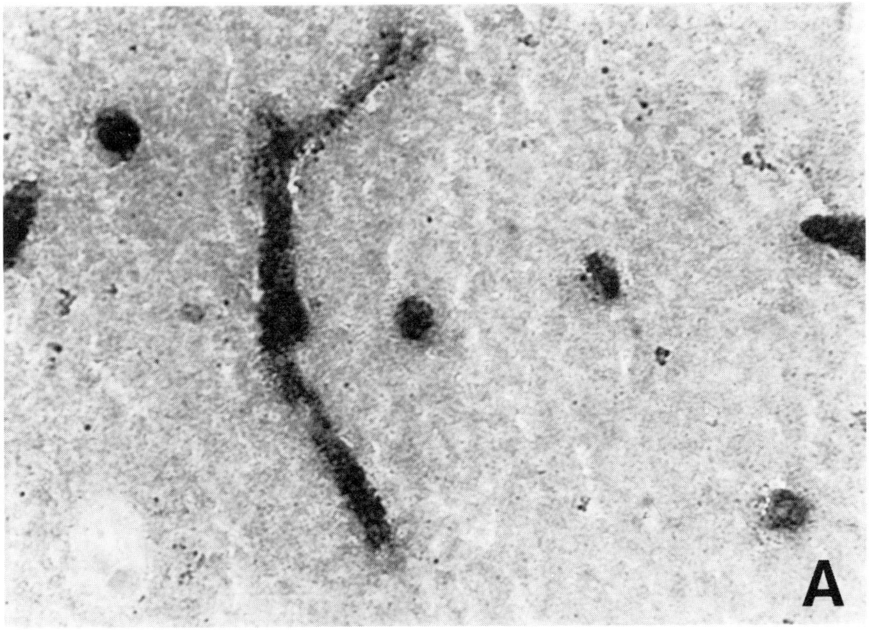
A

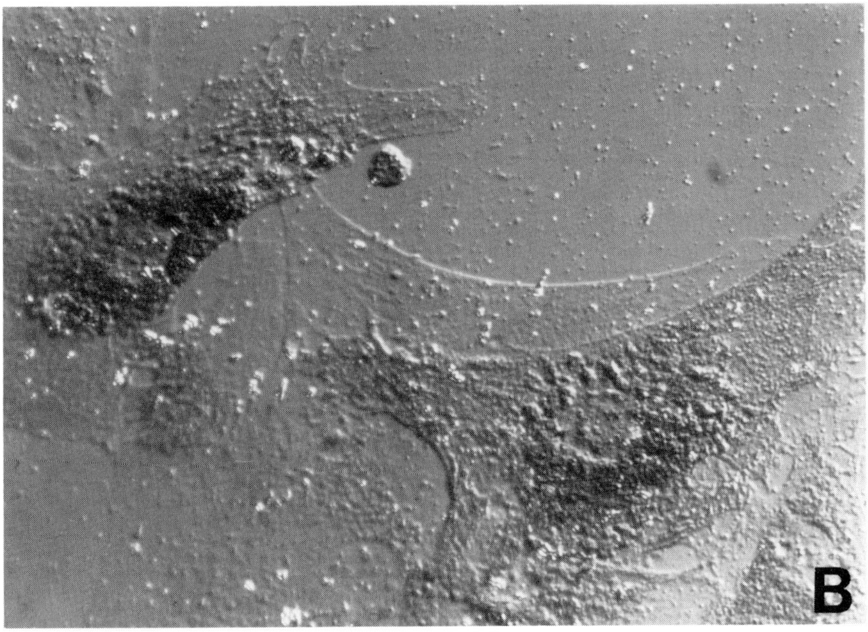
B

plastic dishes or aortic cells on either plastic or native collagen. Treatment of capillary and aortic cells with TAF followed by extensive washings with normal growth medium and subsequent culture of these cells, resulted in the stimulation of capillary but not aortic cells. Supernatants from TAF-treated capillary cells lost their ability to stimulate fresh capillary cells. In contrast, supernatant from TAF-treated aortic cells retained their ability to stimulate fresh capillary cells (a detailed account of these and the following experiments will be published elsewhere).

Most lectins bound similarly to BAEC and BCEC and the binding was not affected by the pretreatments with TAF. However, the bindings of ricin was significantly inhibited by TAF pretreatment of capillary but not aortic endothelial cells (fig. 6). The sugar specificity for ricin is β-D-galactose. Treatment of capillary cells with unlabelled D-galactose abolished more than 90 % of the binding.

BCEC that had attached overnight to collagen gels when exposed to different concentrations of TAF showed that one minute after TAF treatment there was a decrease in the percentage of fusiform and an increase in polygonal cells. The proportion of spherical cells was not significantly altered by the TAF treatment (fig. 8).

Newly trypsinized and replated endothelial cells were usually either blebbed or microvillous. BCEC consistently had lower percentages of blebbed cells and higher percentages of microvillous cells than did BAEC. Thus, there seems to be an intrinsic difference between BCEC and BAEC in their reaction to trypsinisation which may or may not be related to their origin. The interesting feature of trypsinized cells at 1 minute after plating (i. e. 1 minute after addition of TAF) was that both BCEC and BAEC treated with TAF showed fewer blebs and more microvilli than untreated BCEC and BAEC. This was true for six of the seven experiments performed on the two cell types. If this was an effect of TAF on the surface morphology of spherical cells, then it was a transient one, as it had disappeared by 5 minutes after replating.

The percentages of rounded BCEC or BAEC with blebs fell gradually with time but never reached less than 6.7 % in BCEC and less than 19.8 % in BAEC by 90 minutes after replating. On the other hand, the percentages of rounded microvillous cells fell more rapidly, as low as 0.4 % in BCEC and 0.0 % in BAEC by 90 minutes after replating.

TAF treatment had no effect on the fall in percentages of blebbed cells nor on the fall in percentages of BAEC with microvilli. However, in the case of BCEC with microvilli, the maximum percentages of cells with

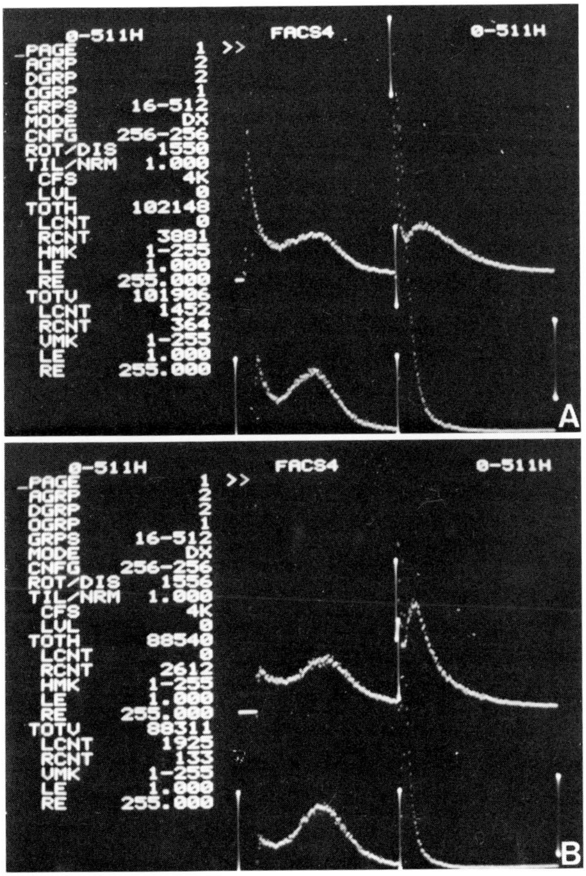

Fig. 6. The effect of TAF and D-galactose on the binding of fluorescein-labelled ricin to capillary endothelial cells was analysed by flow cytometry. The profiles show the size distribution of the endothelial cells sampled (approximately 10^5 cells) on the left and the distribution of fluorescence intensity on the right. Each of the four samples exhibited a similar size distribution. Endothelial cells incubated with ricin alone showed peaks of increased fluorescence due to the ricin binding (top A and top B). Pretreatment of the cells with ricin in the presence of TAF and D-galactose almost completely abolished the ricin binding (bottom A and bottom B respectively). Other lectins (Lens culinaris, Concavalin A, peanut, wheatgerm, succinylated wheatgerm, Helix pomatia, Vicia vellosa, Sophoria japonica, Wisteria floribunda, Soyabean, Ulex europeus, Bandeiraea simplicifolia, Phythohemagglutinin [Erythroglutinin portion of PHA], Dolichos biflorus, Phytohemagglutinin [mixture of E and L PHA] and Pisum sativum) were also used but their binding was not influenced by pretreatment with TAF.

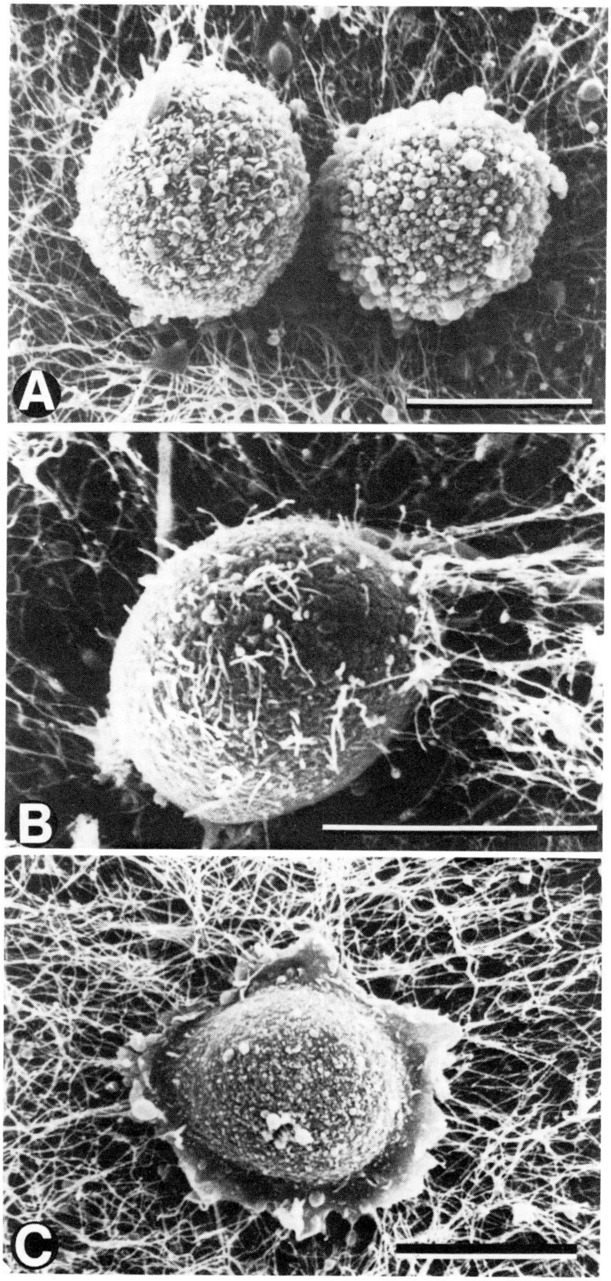
A
B
C

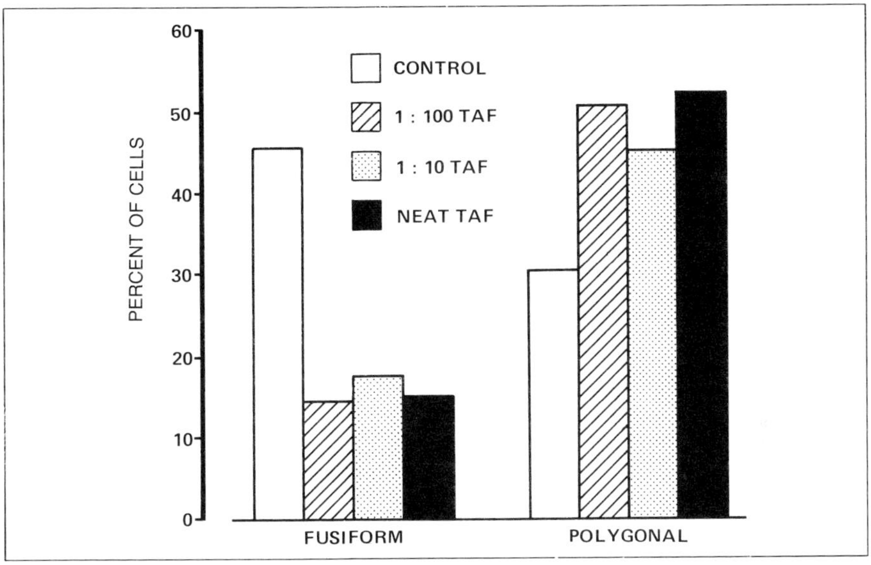

Fig. 8. The effect of TAF treatment on the light microscope morphology of capillary cells which were allowed to attach to collagen gel overnight.

microvilli occurred at 1 minute with TAF and 5 minutes in controls. Thus it appeared that in three out of four experiments with BCEC, microvillous cells peaked earlier with TAF (at 1 minute) than if untreated (peaking at 5 minutes). This may or may not be related to the action of TAF on BCEC.

Flattening down on to a native collagen substratum was considered in two stages, the round to semi-flattened transition and the semi-flattened to flattened transition.

Semi-flattened cells had usually peaked before 90 min in BCEC but never in BAEC. This implied that BCEC were quicker in passing through the semi-flattened stage to flatten completely than were BAEC. TAF treatment neither retarded nor speeded up the transitions from round-semi-flattened or from semi-flattened to flattened in BCEC or BAEC. When the percentages of semi-flattened and flattened cells were com-

---

Fig. 7. Scanning EM study to examine the effect of TAF on the morphology of tissue cultured capillary and aortic endothelial cells (only photomicrographs of capillary cells are shown here). *A* Two spherical blebbed cells (× 6,000); *B* a spherical cell with microvilli (× 9,000); *C* a semi-flattened cell with microvilli and a ruffled peripheral cytoplasm (× 5,800). Bars represent 5 μm.

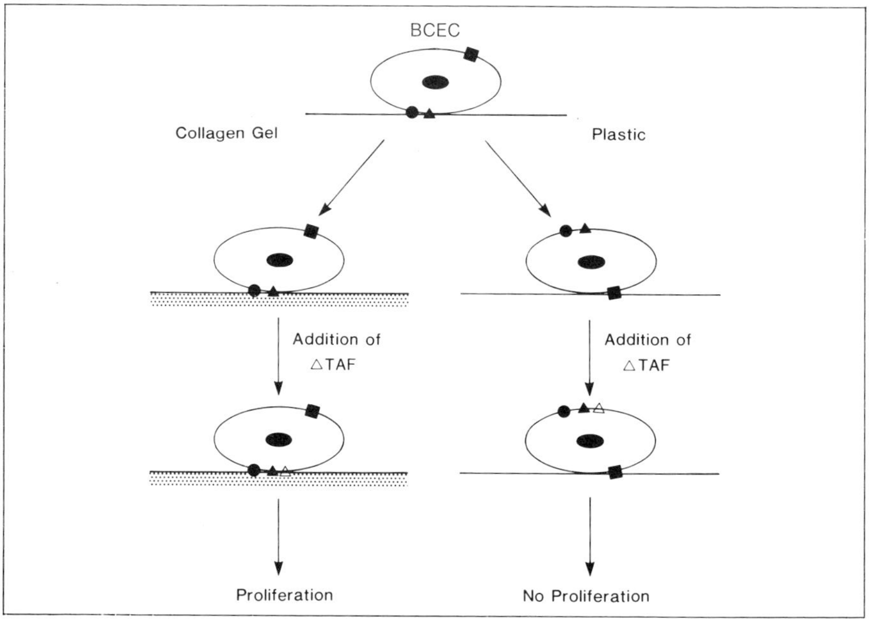

Fig. 9. A diagrammatic model to explain the stimulation of capillary cells and the requirement for a collagen substratum.

bined, analysis of the results showed that TAF had no visible effect on the rates of flattening in either BCEC or BAEC.

TAF had no effect on the levels of cell-associated or secreted elastase-like activity. BCEC did not secrete trypsin-like, chymotrypsin-like, or collagenase activities nor did the addition of TAF stimulate the secretion of any of these enzymes by BCEC.

Discussion

Our previous and present studies have demonstrated a number of differences between capillary and aortic endothelial cells in their response to TAF [10]. Preliminary data also indicate that there is a structural difference in angiogenesis factors from tumours, retina and synovial fluid of patients with rheumatoid arthritis (see below).

Proliferation of capillary but not aortic endothelial cells was enhanced by the addition of TAF and then only when the BCEC were growing on a native collagen gel. It is also apparent that TAF binds to BCEC although this binding in itself is insufficient to induce BCEC proliferation. The latter can only be stimulated by cell attachment to the collagen substratum. The mechanisms by which TAF stimulates BCEC and the reasons for the necessity for native collagen gel are not known. It is possible that cell surface receptors for TAF and the collagen substratum may act together to stimulate cell proliferation. To speculate, it would seem that BCEC possess receptor sites, one for TAF, one for attachment to collagen and one for attachment to the fibronectin coating on plastic tissue culture dishes. It is tempting to suggest that the receptor sites for TAF and collagen are sufficiently close to enable the occupied sites to interact and so stimulate BCEC proliferation, whereas sites for TAF and fibronectin are remote, hence site interactions do not occur (fig. 9).

The reasons for the failure of aortic endothelial cells to respond to TAF is not clear either. This was not a result of inactivation of TAF by aortic cells as TAF pre-exposed to BAEC retained the capacity to stimulate proliferation of fresh BCEC. Rather, the lack of response probably reflected a failure of aortic endothelial cells to bind TAF. That endothelial cells from different sites exhibit metabolic differences has been shown by various authors (for a review see [10]). Some of the important ones between aortic and capillary cells observed by us are presented in table I. We have also observed significant differences between normal brain and brain tumour capillary endothelial cells. The number of W-P bodies in normal brain endothelial cells is very much lower compared with those of

Table I. Some characteristics of tissue cultured bovine capillary and aortic endothelial cells

Endothelial marker	Source of endothelial cells	
	Bovine brain capillary	Bovine aorta
1. Morphology	Atypical (almost fibroblastic)	Typical *(Cobblestone)*
2. Factor VIII	++	++
3. Alkaline phosphatase	++	+
4. γ-glutamyl transpeptidase	++	–
5. Weibel-Palade bodies	Rare*	More frequent
6. Angiotensin converting enzyme	Negligible	++
7. Prostacyclin	Negligible	++

– Absent; + present; ++ marked

* Normal human brain capillary endothelial cells contain very few W-P bodies but the latter are numerous in endothelial cells of human brain tumour capillaries.

brain tumours (*Kumar et al.* [11]). Our unpublished data has also shown that normal synovial capillary endothelial cells contain significantly fewer numbers of W-P bodies than biopsy specimens from rheumatoid arthritis patients. *Gospodarowicz et al.* [6] demonstrated that human umbilical vein endothelial cells possessed receptors for epithelial growth factor, whereas bovine umbilical vein and aortic endothelial cells did not. *Bar et al.* [3] reported that the human umbilical artery endothelial cells bound twice as much insulin as did the venous cells, although both these cell types bound less multiplication stimulating activity than did endothelial cells cultured from primary arteries and veins [2]. Folkman's group has also reported differences in the migration of capillary and aortic endothelial cells in response to an angiogenesis factor. They found that capillary but not aortic endothelial cells migrated and proliferated in response to angiogenesis factor [17, 18]. Recently, we have studied the effect of a migration factor on the order of the plasma membrane of bovine aortic and capillary endothelial cells by Electron Spin Resonance technique, using the lipophilic spin label 5-doxyl stearic acid [*Dodd and Kumar*, J. Cell Sci., in press, 1984]. Exposure of BCEC membranes to a migration factor caused a reduction of membrane order but had no detectable effect on membranes of aortic endothelial cells. This probably reflects differences in lipid composition aortic cell membranes being inherently less rigid than those of capillaries. These results are consistent with the above-mentioned findings of *Zetter* [17] that capillary endothelial cells move slowly in the absence of TAF and rapidly in its presence; aortic cells move more rapidly irrespective of treatment of TAF.

The inhibition of ricin binding by TAF to capillary cells is interesting. *Badley et al.* [1] found that substratum adhesion sites of tissue cultured fibroblasts contain binding sites for ricin. Adhesion and spreading in anchorage dependent cells are necessary for cell division to occur. Therefore, we postulated that a mechanism of action of TAF might be to promote adhesion of capillary cells to the substratum. In view of the effects of epidermal growth factor and nerve growth factor in increasing cell adhesion [4, 9] the results of our scanning EM study are rather disappointing in that TAF did not increase the rate of flattening of capillary or aortic endothelial cells on to a collagen substratum. The use of freshly trypsinized cells in the attachment study should not have interferred in the adhesion process because it has been shown that the glycoproteins involved in adhesion are resistant to enzymatic cleavage by trypsin [16].

Capillary and aortic endothelial cells were found to become blebbed or microvillous to different extents after trypsinization. *Porter et al.* [15] and *Harrison and Allen* [7] suggested that microvilli and blebs were associated with each recognizable stage of the cell cycle. Since both capillary and aortic cells were asynchronous populations, it is difficult to evaluate their morphology or change in morphology in terms of cell cycle. Experiments using a cell sorter which provides G_1 or S-G_2 cell populations might reveal an answer to this problem.

An early effect on the cytoskeleton was observed in light microscopic experiments. After TAF-treatment there was significant increase in the percentages of polygonal cells while the percentages of fusiform cells decreased. Work is now in progress to examine the effect of TAF on the cytoskeleton of capillary cells using cytoskeletal depolymerisers (colcemid, cytochalasin B) and stabilizers (phalloidin, taxol).

The fact that TAF had no detectable effect on the levels of cell associated or secreted elastase-like activity needs comment. *Hatcher et al.* [8] have shown that a relationship exists between cell surface proteinase activity and the rate of proliferation of various cell types. The addition of TAF to capillary cells would, therefore, be expected to enhance the levels of proteinases. This apparent lack of stimulation of endopeptidase activity by TAF warrants closer examination of the presumptive role of these enzymes in angiogenesis.

Our preliminary data indicate that purified angiogenesis factors from tumour, bovine retina, and synovial fluid of patients with rheumatic diseases, although indistinguishable in their physical and chemical properties, behave differently as regards their interference with ricin binding to endothelial cells. It needs to be emphasized that these data need confirmation. However, they showed that tumour angiogenesis factor but not angiogenesis factors from retina or synovial fluid inhibited the binding of ricin. We believe this is an important observation with profound implications. It is entirely conceivable that apparently similar angiogenesis factors have subtle structural differences which might have a pivotal role in their mechanism of action. In a recent paper [13] we isolated an angiogenesis factor from human myocardial infarctions which appeared to be similar to one isolated from solid tumours. Neovascularization in response to tumour angiogenesis factor occurs as an ingrowth of new capillaries from the surrounding host tissues. However, in myocardial infarction there are two mechanisms by which the damage of an infarct can be repaired. The first is an immediate response leading to enlargement of collateral (large) blood vessels and it is the second much later event (capillary ingrowth in dead

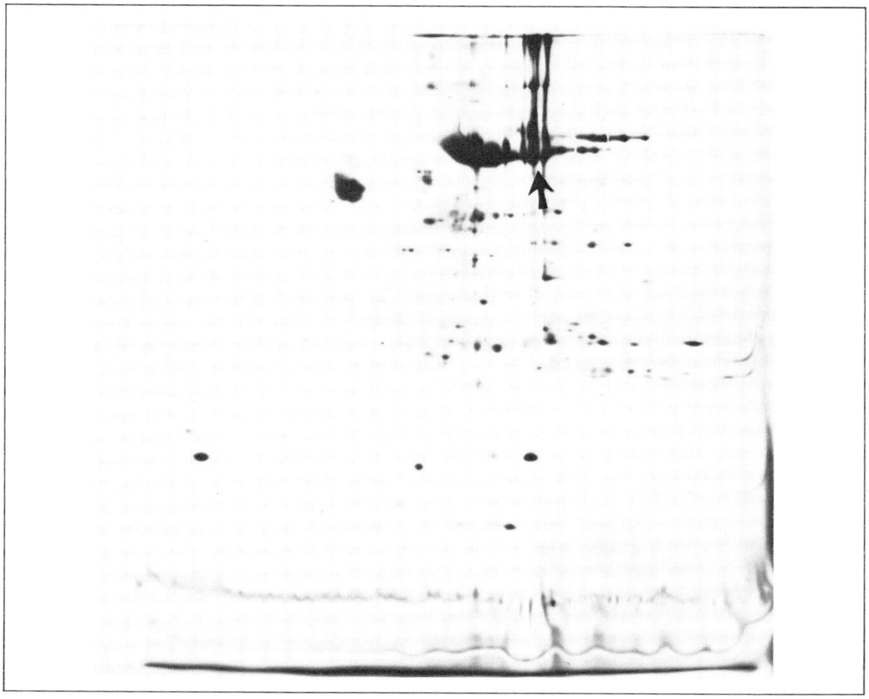

Fig. 10. Two-dimensional gel electrophoresis of a semi-pure TAF preparation (G-100 Sephadex eluted from a single peak). The high number of spots is rather surprising; one (arrowed) represents albumin which must have originated from serum.

tissue), which is comparable to tumour-induced angiogenesis. Whether or not a single angiogenesis factor can cause both collateral enlargement and the growth of capillary sprouts or whether they are induced by two different factors is not known. When answers to such questions are available we should begin to understand the process of angiogenesis and the relevance of in vitro assays in the study of angiogenesis.

The value of the present in vitro assay is severely limited by lack of a capillary endothelial cell line. The use of primary cultures necessitates standardization with every batch of new culture. A library of capillary endothelial cell lines adapted to grow in serum-free medium would be needed to develop an in vitro assay. The tissue source of angiogenesis factor is another limitation. Studies thus far have either used extracts of autopsy material, biopsy specimens or serum-free washes of tissue-cultured primary cell cultures or cell lines. The use of such materials has not

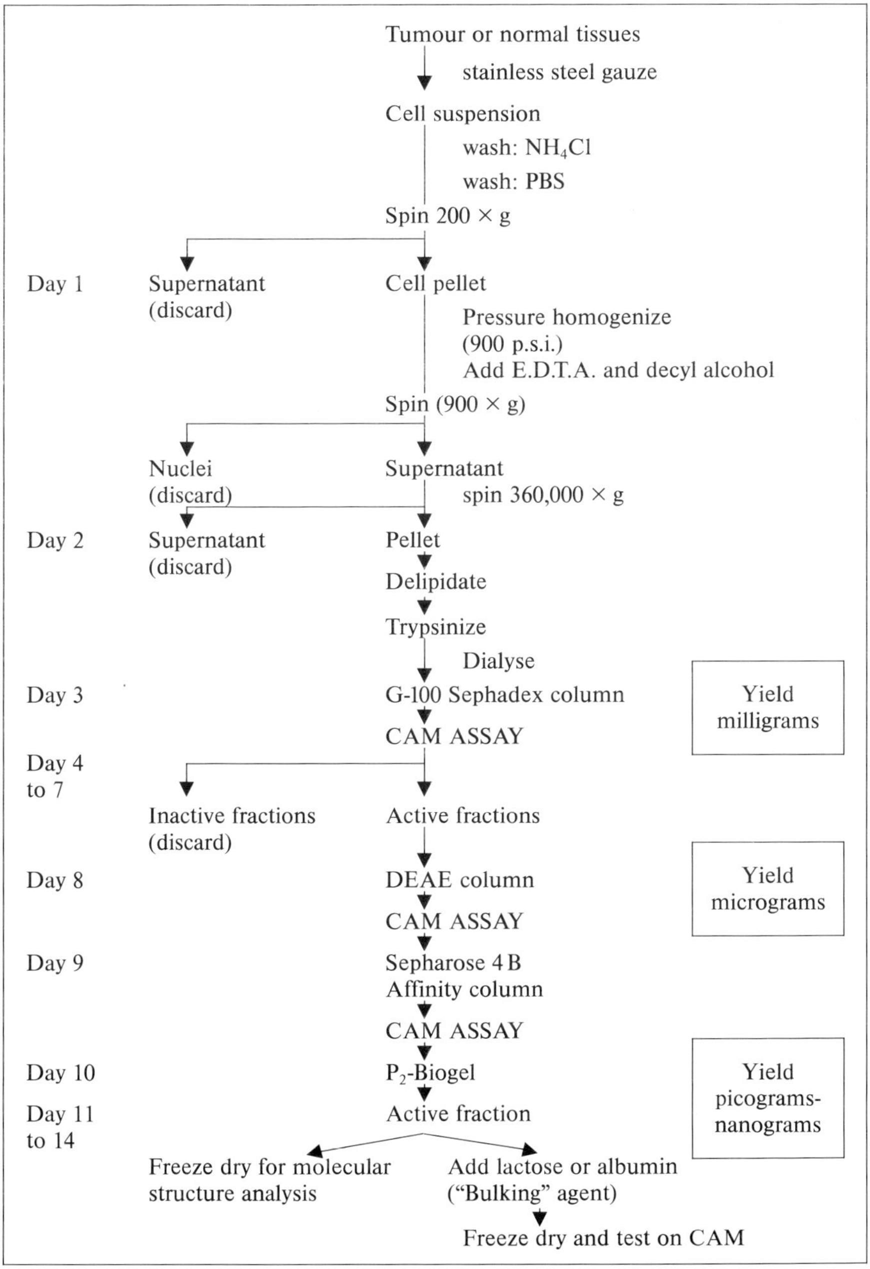

Fig. 11. Flow diagram of the extraction and purification procedure.

allowed an answer to the following two questions: (1) the cell type in which the angiogenic activity originated, (2) is the angiogenic factor endogenous or acquired by the cells from the extracellular environment, e.g. serum. Figure 10 illustrates this point wherein active material obtained after G-100 Sephadex column purification (see flow diagram, figure 11) on two-dimensional gel electrophoresis, showed a number of protein spots. One (arrowed) corresponds to albumin, which must have originated from serum. Recently we have isolated a human lung tumour cell line and have adapted it to grow in serum-free medium for over 12 months [12]. The culture medium contained an angiogenesis factor. Thus for the first time a standard source of TAF, free from other cell types and serum contamination, has become available which will simplify the purification and eventual characterization of the active molecules.

Summary

Our previous and present studies have demonstrated a number of differences between bovine capillary and aortic endothelial cells in their morphology, enzyme content and response to TAF. Preliminary results also indicate that purified low molecular weight, angiogenesis factors from tumours, synovial fluid of patients with joint diseases and normal retina, although indistinguishable from one another, might have subtle structural differences.

Tumour angiogenesis factor, which was angiogenic in vivo also enhanced proliferation of bovine capillary but not aortic endothelial cells in vitro. This difference between large and micro-vessel derived endothelial cells was due to non-binding of TAF and was not a result of inactivation of TAF by aortic cells. Pretreatment of capillary cells with TAF significantly inhibited their staining with fluorescein labelled ricin. Scanning EM studies showed that the addition of TAF had no detectable effect on the morphology of freshly trypsinized capillary or aortic endothelial cells. In contrast, an early effect on cytoskeleton of capillary endothelial cells that had attached overnight to collagen gels was observed using light microscopy. Neither capillary nor aortic endothelial cells secreted significant amounts of proteolytic enzymes and the addition of TAF did not modulate these levels. Difficulties and limitations of the present in vitro assay as a model for angiogenesis are discussed.

References

1 Badley, R. A.; Lloyd, C. W.; Woods, A.; Carruthers, L.; Allcock, C.; Rees, D. A.: Mechanisms of cellular adhesion. III Preparation and preliminary characterisation of adhesions. Exp. Cell Res. *117:* 231–234 (1978).

2 Bar, R. S.; Peacock, M. L.; Rechler, M. M.; Goldsmith, J.: Presence of multiplication-stimulating activity (MSA) receptors in human and bovine endothelium. Clin. Res. *28:* 255A (1980a).

3 Bar, R. S.; Peacock, M. L.; Spanheimer, R. G.; Veenstra, R.; Hoak, J. C.: Differential

binding of insulin to human arterial and venous endothelial cells in primary culture. Diabetes *29:* 991–995 (1980b).
4 Berliner, J. A.; Aharanov, A.; Pruss, R.: Cell surface changes associated with EGF and insulin-induced adhesion. Exp. Cell. Res. *133:* 227–235 (1981).
5 DeBault, L. E.: γ-glutamyl-transpeptidase induction mediated by glial foot process to endothelium contact in co-culture. Brain Res. *220:* 432–435 (1981).
6 Gospodarowicz, D.; Brown, K. D.; Birdwell, C. R.; Zetter, B. R.: Control of proliferation of human vascular endothelial cells: characterization of the response of human umbilical vein endothelial cell to fibroblast growth factor, epidermal growth factor and thrombin. J. Cell Biol. *77:* 774–788 (1978).
7 Harrison, C. J.; Allan, T. D.: Cell surface morphology after trypsinization depends on initial cell shape. Differentiation *15:* 61–66 (1979).
8 Hatcher, V. B.; Wertheim, M. S.; Rhee, C. Y.; Tsien, G.; Burk, P. G.: Relationship between cell surface protease activity and doubling time in various normal and transformed cells. Biochim. biophys. Acta *451:* 499–510 (1976).
9 Huff, K.; End, D.; Guroff, G.: Nerve growth factor-induced alteration in the response of PC12 pheochromocytoma cells to epidermal growth factor. J. Cell Biol. *88:* 189–198 (1981).
10 Keegan, A.; Hill, C.; Kumar, S.; Phillips, P.; Schor, A.; Weiss, J.: Purified tumour angiogenesis factor enhances proliferation of capillary, but not aortic, endothelial cells in vitro. J. Cell Sci. *55:* 261–76 (1982).
11 Kumar, P.; Kumar, S.; Marsden, H. B.; Lynch, P. G.; Earnshaw, E.: Weibel-Palade bodies in endothelial cells as a marker for angiogenesis in brain tumours. Cancer Res. *40:* 2010–2019 (1980).
12 Kumar, S.; West, D.; Daniel, M.; Hancock, A.; Carr, T.: Human lung tumour cell line adapted to grow in serum-free medium secretes angiogenesis factor. Int. J. Cancer *32:* 461–464 (1983).
13 Kumar, S.; West, D.; Shahabuddin, S.; Arnold, F.; Haboubi, N.; Reid, H.; Carr, T.: Angiogenesis factor from human myocardial infarcts. Lancet *ii:* 364–368 (1983).
14 Phillips, P.; Kumar, P.; Kumar, S.; Waghe, M.: Isolation and characterization of endothelial cells from rat and cow brain white matter. J. Anat. *129:* 261–272 (1979).
15 Porter, K. R.; Puck, T. T.; Hsie, A. W.; Kelley, D.: An electron microscope study of the effects of dibutyryl cyclic AMP on Chinese Hamster ovary cells. Cell *2:* 145–162 (1974).
16 Tarone, G.; Galletto, G.; Prat, M.; Comoglie, P. M.: Cell-surface molecules and fibronectin-mediated cell adhesion: effect of proteolytic digestion of membrane protein. J. Cell Biol. *94:* 179–186 (1982).
17 Zetter, B. R.: Migration of capillary endothelial cells is stimulated by tumour-derived factors. Nature, Lond. *285:* 41–43 (1980).
18 Zetter, B. R.; Azizkhan, R. G.; Azizkhan, J. C.; Brouty-Boye, D.; Folkman, J.; Haudenchild, C. C.; Klagsburn, M.; Potash, R.; Scheiner, C. J.: Normal and tumour-derived factors that modulate endothelial cell growth and migration; in Bing, Rosenbaum (eds.), Plasma and Cellular Modulatory Proteins, pp. 59–73 (Center for Blood Research Inc., Boston, Mass. 1981).

Dr. S. Kumar, Christie Hospital, Manchester M 20 9 BX (England)

Prog. appl. Microcirc., vol. 4, pp. 76–87 (Karger, Basel 1984)

Low Molecular Weight Angiogenesis Factor: A Growth Factor not Unique to Tumours which Activates Procollagenase

J. B. Weiss[1], *S. F. Elstow*[1], *C. R. Hill*[3], *B. McLaughlin*[1], *E. M. Davidson*[1], *A. Schor*[4], *S. R. Ayad*[2]

[1] Department of Rheumatology
[2] Department of Biochemistry, University of Manchester
[3] Cell-Tech, Slough, Berks
[4] CRC Department of Medical Oncology, Christie Hospital and Holt Radium Institute, Manchester, England

Introduction

An association between tumour development and neovascularisation has been recognized for nearly forty years [2]. This association was confirmed when *Greenblatt and Shubik* [15] were able to demonstrate that a tumour, separated from its host by a thin porous membrane, could still elicit a blood vessel response in the host. A natural inference from this work was that neovascularisation could be induced by a diffusible factor secreted from the tumour cell. Such a humoral factor was isolated, in a crude form by *Folkman* and co-workers in 1971 [12] and said to be a large molecule ($M_r \simeq 100{,}000$) consisting of protein, nucleic acid and carbohydrate. Later work, using the same tumour source (rat Walker 256) showed that the crude material described by *Folkman* when applied to a DEAE column was a mixture of many proteins and glycoproteins [27]. The angiogenic activity was not invariably associated with any particular peak eluting from the DEAE column, leading to the supposition that the proteins were non-specific carriers for a smaller molecule. That this was indeed the case was confirmed by these workers who employed an affinity chromatography column which utilised an antibody to the crude starting material described by *Folkman et al.* raised in rabbits [21]. By this means a potent low molecular weight ($M_r \simeq 300-600$) angiogenic factor was obtained [27].

This factor has been shown to stimulate proliferation of cultured capillary endothelial cells grown on collagen gels [17, 23], as well as to induce neovascularisation in the chick chorioallantoic membrane test. Both these activities being observed using amounts of the factor too low to be measureable by weight or U. V. absorption. Confirmation of the presence of a low molecular weight angiogenesis factor from rat Walker 256 tumour cells came from *Fenselau et al.* [11] who used ethanol extraction of the homogenate of the solid tumour. Other workers [20] also demonstrated low molecular weight activity in homogenates of rat Walker 256 tumour cells grown in culture conditions. Although the crude high molecular weight factor had been called tumour angiogenesis factor by *Folkman et al.* [12] it was always unlikely that a tumour cell would produce a unique substance. It was not therefore surprising when evidence for angiogenic factors from non-tumour sources was obtained. The first of these non-tumour low molecular weight angiogenic factors to be described was obtained from synovial fluids [5, 6] and the next from feline retinas [19], crude extracts of which had previously been shown in the rabbit corneal assay to stimulate growth of the limbal vessels [18]. For this reason *Weiss* et al. [27] called the low molecular weight material endothelial cell stimulating angiogenesis factor (ESAF), rather than tumour angiogenesis factor (TAF).

Some doubt was cast on the molecular size of the retinal ESAF by workers who were only able to demonstrate a high molecular weight angiogenic factor in bovine retinas which consisted of two components of molecular mass 70,000 and 50,000 by gel electrophoresis [7, 14].

The work we now report was undertaken to establish the nature of the bovine retinal factor, the ability of a cloned lymphoma cell line to produce the same factor in tissue culture conditions and for it to be obtained from both sources without resorting to affinity chromatography. At the same time a possible mechanism for the action of ESAF will be described which may enable a quantitative assay for the factor to be set up.

Although the chick chorioallantoic membrane (CAM) as an angiogenesis test has been criticised [22] we believe it to be adequate. However, we have used the capillary endothelial cell stimulation assay as a confirmatory test in all cases.

Methods

Preparation of ESAF from Culture Medium of Cloned Mouse Lymphoma Cells (P388–F36 cell line)

The cloned mouse lymphoma cell line and its culture conditions were first described by *Ayad and Delinassios* in 1974 [3]. The culture medium was withdrawn from lymphoma cells in spinner culture at regular intervals. The cells were removed by centrifugation and the culture medium extracted according to the method of reference [16].

Preparation of ESAF from Bovine Retinas

For extraction of angiogenic activity from bovine retinas two methods were employed. The first was that of *Glaser et al.* [14] which used leaching with isotonic saline and the second according to *Kissun et al.* [19] who used homogenisation. Extracts prepared by both methods were further purified by ion exchange chromatography using DEAE cellulose [16] and gel filtration on Biogel P2 in 10% (v/v) propan-2-ol/water. Some extracts were freeze dried and incubated with 10% (v/v) propan-2-ol/water at 56° for 30 minutes or with 100% ethanol at 4° for 2 hours in order to release low molecular weight angiogenic activity.

Chick Chorioallantoic Membrane (CAM) Test for Angiogenic Activity

Extracts of retina and lymphoma cell culture medium were tested for angiogenesis at various stages of purification using the CAM test [12]. Samples were applied to the membrane either directly from ion exchange columns in salt or were bulked in lactose after gel filtration. On some occasions both these test samples were incorporated into Elvax (Ethylene/Vinyl acetate copolymer – Aldrich Chemical Company, England). All samples were sterile filtered prior to testing and were applied to 3 eggs for each test. Angiogenic responses were assessed four and five days after applying material and graded as zero response, weakly positive and strongly positive.

Preparation of Human Skin Fibroblast Procollagenase

Procollagenase was prepared from serum-free human skin fibroblast conditioned medium using a modification of the method of *Stricklin et al.* [25]. The enzyme was capable of being activated by Mersalyl and the metalloproteinase nature of the preparation was confirmed by the EDTA inhibition of the Mersalyl activated proenzyme. Analysis of the degradation products from the enzyme-catalysed hydrolysis of Type I collagen by SDS polyacrylamide gel electrophoresis revealed the quarter and three quarter collagen fragments characteristic of the action of mammalian collagenase [26].

Procollagenase Activation Assay

Type I and Type III collagen substrates were prepared from foetal calf skin [9] and labelled with ^{3}H acetic anhydride according to reference [13]. Purified fractions of the angiogenic factor prepared from extracts from lymphoma cell culture medium were assayed for their ability to activate procollagenase. Only those fractions which were freely dialysable and of low molecular weight (below 600 M_r) as determined by gel filtration were used. This was to ensure that there was no contamination of the fraction with proteolytic enzymes. The collagenase assay system was as described in reference [28]. Briefly the labelled collagen is dissolved in a Tris/HCl buffer and ESAF or other substances which activate the procollagenase or degrade the collagen, are added together with the procollagenase. After incubation for 16 hours at 36.5° C the undegraded collagen is precipitated and the radioactivity of the collagen fragments in the supernatant gives a measure of the amount of degraded collagen and corresponds to the degree of activation of the enzyme. ESAF was used both alone and together with heparin, in order to determine whether there was any potentiating effect.

Stimulation of Endothelial Cell Growth by ESAF

Bovine retinal and brain white matter capillary endothelial cells were grown on collagen and gelatin substrata according to the method of *Schor et al.* [23].

Results

Angiogenesis Factor from Culture Medium from Cloned Mouse Lymphoma Cells

The CAM test was used to monitor the presence of angiogenic activity during the purification of the mouse lymphoma culture fluid. Angiogenic activity was found bound to DEAE cellulose (fig. 1). When the active bound fraction was desalted on Biogel P2 the activity eluted with a volume corresponding to a molecular mass of between 300 and 600 M_r (fig. 2). Control medium extracted under the same conditions gave similar elution profiles indicating that the absorbance was due to components other than ESAF. The low molecular weight material was positive in the CAM test (fig. 3a). The amount of material which gave a consistently

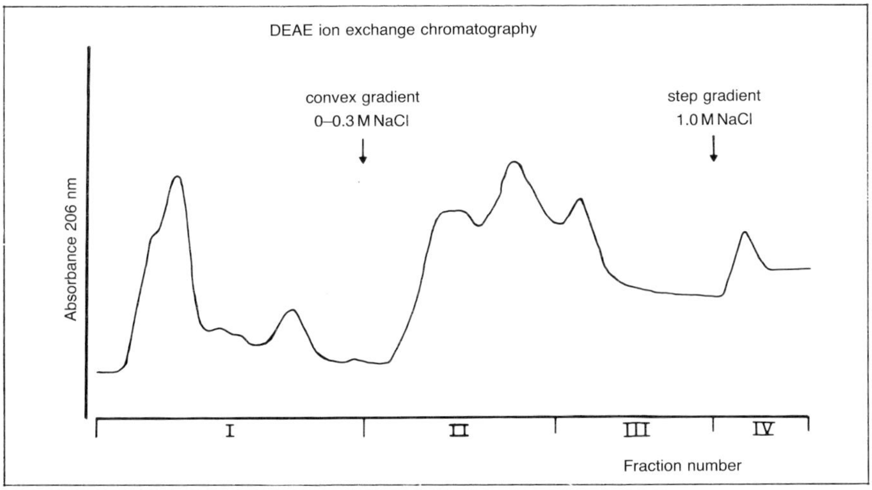

Fig. 1. DEAE Ion exchange chromatography of extracted culture medium of mouse lymphoma cells. The active material eluted in fraction II (Column 12 × 1 cm, flow rate 1ml/minute eluting solution 50 mM ammonium bicarbonate pH 7.9).

positive result on 3 eggs was derived from 1×10^7 lymphoma cells or about 3–10 ml culture medium.

Bovine Retinal Angiogenic Factor

Both methods of preparation of angiogenic factor from bovine retinas, leaching or homogenisation, gave positive results for angiogenesis in the CAM test (fig. 3b). When these extracts were subjected to alcohol treatment either ethanol extraction at 4° or 10% propan-2-ol at 56° a low molecular weight angiogenic factor could be obtained. No angiogenic activity then remained associated with the high molecular weight material.

Prior to alcohol treatment the high molecular weight material was applied to SDS polyacrylamide gel electrophoresis. Two bands were obtained which were the same as those described by *D'Amore et al.* [7]. Bovine serum albumin applied in the same system gave identical bands.

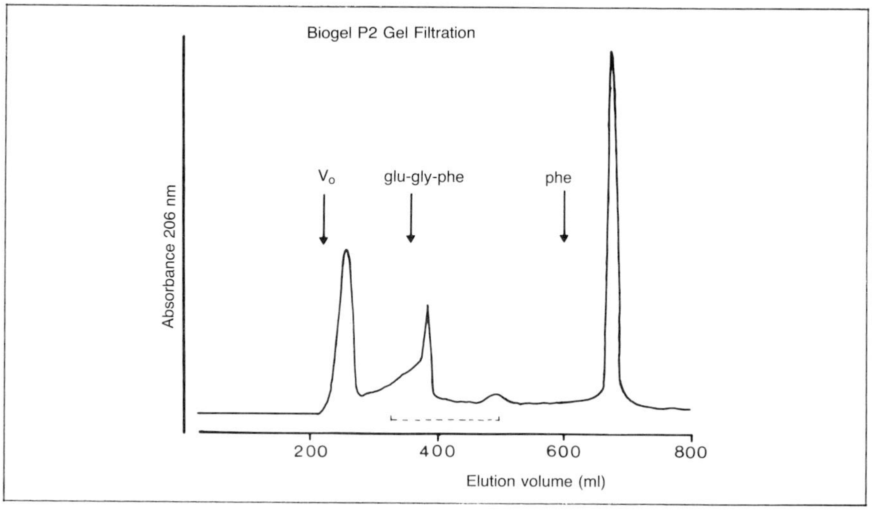

Fig. 2. Biogel P2 gel filtration of fraction II from the DEAE separation shown in figure 1. The tripeptide glutamylglycylphenylalanine (Glu-Gly-Phe) was used as a molecular weight marker (M_r 357) together with phenylalanine alone (M_r 165). The active component eluted with volume (300–500 ml) corresponding to a molecular mass of 300–600. (Column 4.4 cm × 44 cm, flow rate 0.75 ml/minute, eluting solution 10% (v/v) propan-2-ol in water). (The use of propan-2-ol in the eluting solution reduces the exclusion limits of the column from 2,000 to 1,000 M_r).

Capillary Endothelial Cell Stimulation

The low molecular weight bovine retinal angiogenic factor was used to stimulate proliferation of bovine capillary endothelial cells from both retina and white matter grown on a collagen substratum. The growth

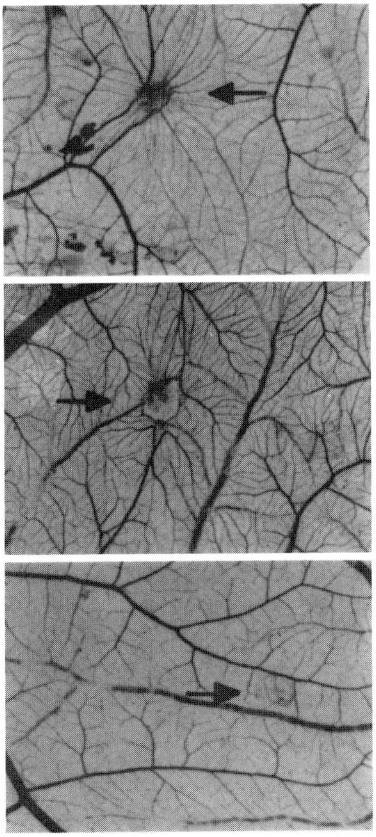

a

b

c

Fig. 3. Chick chorioallantoic membrane (CAM) test results for retinal and lymphoma ESAF. The sample was applied in either an Elvax pellet (*b, c*) or directly in salt (*a*). Key: *a* ESAF extracted from lymphoma cell medium applied in salt (100 μg): The amount applied corresponded to 10^7 lymphoma cells; *b* retinal ESAF applied in lactose (100 μg) incorporated into an Elvax pellet (total weight not exceeding 300 μg); *c* Control. As in (*b*) but without ESAF.

Arrow indicates the position of applied material. *Note:* Salt alone did not give a positive response.

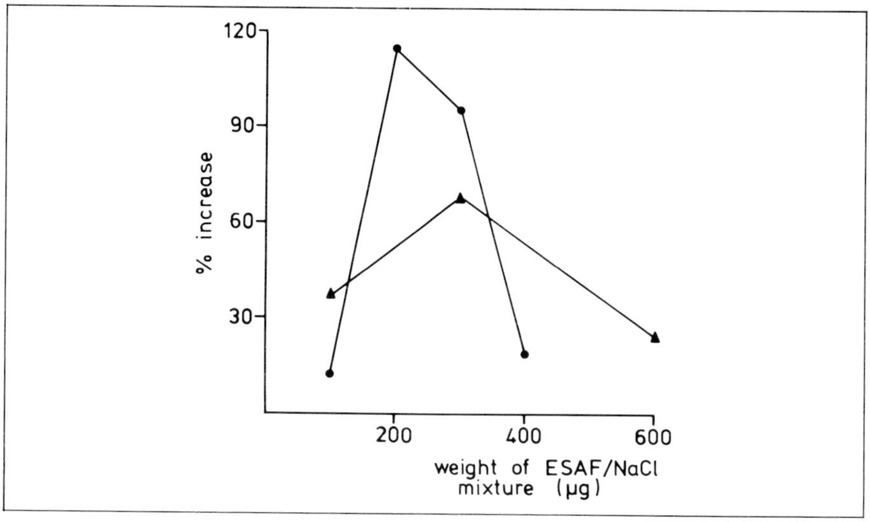

Fig. 4. Effect of bovine retinal ESAF on capillary endothelial cells. ●——● bovine white matter endothelial cells ▲——▲ bovine retinal endothelial cells. All cells were grown on a Type I collagen substratum. Results are expressed as percentage increase in cell numbers over controls grown over a four-day-period. Each point is the mean of triplicate cultures. ESAF was applied bulked in salt and weights represent the amount of salt ad-mixed with ESAF. The amount of ESAF itself is too small for accurate measurement.

curves obtained for both types of cells are shown in figure 4. Both growth response curves were bell shaped which is characteristic of growth factors. When gelatin was used as an alternative substratum, no additional proliferation of endothelial cells over the control was observed.

Previously the lymphoma derived ESAF had also been shown to stimulate proliferation of bovine white matter capillary endothelial cells [30].

Procollagenase Activation

Procollagenase can be activated by the organomercurial compound Mersalyl [24]. A comparison between the activation with this compound and with low molecular weight ESAF is shown in table I. It can be seen that ESAF was, in fact, a more effective activator of procollagenase against both substrates than the organomercurial compound. There was no effect of ESAF on either type of collagen alone and ESAF had no effect when added to the fully activated enzyme (table I). Trypsin was used to activate the enzyme prior to addition of ESAF in this latter experiment.

Effect of Heparin on ESAF Activation of Procollagenase

Heparin was added to the procollagenase-collagen incubation mixture either alone or with ESAF. The results can be seen in table II. It can be seen that whilst heparin had virtually no effect, it markedly potentiated the effect of ESAF on the activation of the proenzyme.

Discussion

Chick Chorioallantoic Membrane (CAM) Test for Angiogenic Activity

In our hands the CAM test has proved to be a very reliable testing system for angiogenic activity. We have rarely obtained a false positive, as confirmed by other test systems for angiogenesis, however we do obtain a zero response when the amount of material applied is insufficient. We have used two systems of application, the Elvax pellet and direct application in a lactose or salt carrier. Criticisms of the CAM as a suitable test system for angiogenesis are based on physical rather than biochemical factors affecting the membrane [22]. The size of our Elvax pellets, containing test substances, is of the order of 1 mm^2 with a depth ranging

Table I. Procollagenase activation by ESAF

Incubation mixture	Collagen degraded (μg/16h)	
	Type I	Type III
Enzyme	0.5 ± 0.1	0
ESAF	0.3 ± 0.1	0.2 ± 0.1
Enzyme + ESAF	26.5 ± 0.3	17.8 ± 0.3
Enzyme + Mersalyl	12.1 ± 0.1	15.2 ± 0.1
Enzyme + Trypsin	23.2 ± 0.6	22.3 ± 0.5
Enzyme + Trypsin + ESAF	23.1 ± 0.5	22.5 ± 0.4

Table II. Effect of ESAF and Heparin on the activation of procollagenase

Incubation mixture	Type I collagen degraded (μg/16h)
Enzyme	0
ESAF	0.3 ± 0.1
Heparin	0.2 ± 0.1
Enzyme + ESAF	1.7 ± 0.1
Enzyme + Heparin	0.2 ± 0.1
Enzyme + ESAF + Heparin	4.3 ± 0.1

Note: Smaller quantities of ESAF were used in this experiment than were used in the experiment reported in table I in order to show potentiation.

between 0.2 mm and 0.5 mm. The weight of salt or lactose applied either directly or in Elvax is less than 100 μg. The weight due to the Elvax alone is 100–200 μg. Control tests using Elvax alone or lactose or salt incorporated into Elvax give no response (fig. 3c). Similarly, salt and lactose applied directly gave no response. In order to assess the response of the CAM to angiogenic materials we view the microvasculature of the membrane under 10× magnification at four and five days after applying the test substance. For a positive result the new blood vessel growth must be directed towards the pellet from all sides and must be observed in all of the three eggs used per test. When the low molecular weight material is applied directly a positive response can usually be seen on the fourth day and no change occurs between days four and five, whereas when it is applied incorporated in an Elvax pellet the response on day five may be greater than on day four. This we believe is due to the material being released more slowly. We are aware that any positive response of our test substance may be due to its potentiating effect on angiogenic substance already present and acting in the CAM. For this reason we like to use more than one test system for detecting angiogenic substances.

We have shown that the mouse lymphoma cell line P388–F36 is able to produce low molecular weight ESAF in culture conditions. This ESAF is able to elicit a positive response on the CAM test and is also capable of activating procollagenase. The amounts of ESAF applied to the CAM and used in the procollagenase activation system were equivalent to that produced in four days by 10^7 cells or less. This suggests that the material extracted from so few cells must be quantitatively very small and therefore, by defintion, very potent.

The original description given for the retinal angiogenic factor [7, 14] was of a high molecular weight protein with 2 components of molecular weights 70,000 and 50,000. Subsequently, *Kissun et al.* [19] using an affinity chromatography column, which used an antibody raised to crude tumour angiogenesis factor [21], were able to demonstrate low molecular weight angiogenic activity in cat retina. The data we now present confirms the presence of a low molecular weight angiogenic factor, this time in bovine retina, which can be released from the protein carrier without recourse to affinity chromatography. Furthermore, it seems very probable that the protein bands reported by *D'Amore et al.* [7] as being the bovine retinal angiogenic factor are in fact components of bovine serum albumin.

Carriers for ESAF may be specific or non-specific and albumin is well known as a non-specific, often low affinity, carrier for a number of small molecules in the blood.

We feel we cannot sufficiently stress the fact that on other occasions when we have investigated other angiogenic factors from different tissues, we have only been able to obtain low molecular weight endothelial cell stimulating angiogenesis factor (ESAF) and no activity has remained with the high molecular weight protein [8, 27]. This is exactly what we now report for the bovine retina. All this evidence is consistent with the concept of a carrier system for ESAF and probably explains the wide range of high molecular weights reported in the literature for various angiogenic factors [10].

Activation of procollagenase by ESAF has been reported previously [28] and it has been suggested that this activity may be involved in the ability of ESAF to direct new blood vessel growth towards the source of the factor. Our more recent observation that heparin is able to potentiate this effect [29] appears to confirm this suggested mechanism of action. Heparin has been shown to potentiate the effect of tumour angiogenesis factor on the chick chorioallantoic membrane [4].

It is expected that the composition of the CAM is similar to other embryonic membranes such as amnion and chorion in which the major collagen types present are types I and III [1]. Both these collagens are susceptible to skin fibroblast procollagenase and it would be expected that a procollagenase would be naturally present in the membrane associated with the polymeric forms of these collagens and the activation of the enzyme would contribute to part of the angiogenic response.

A more pragmatic approach to the ability of ESAF to activate procollagenase would be to extend the assay described here for the quantitation of ESAF.

Summary

Endothelial cell stimulating angiogenesis factor (ESAF) which is a low molecular weight material (Mr 300–600) has been isolated from the medium of a cloned mouse lymphoma cell line in culture and from bovine retinas. The factor has been shown to stimulate proliferation of both bovine retinal and bovine white matter capillary endothelial cells grown on a collagen substratum and gives bell shaped dose response curves.

The factor also activates latent or procollagenase from human skin fibroblasts and this effect is potentiated by heparin. It is hoped that this property may be the basis of a quantitative assay for ESAF.

References

1 Abedin, M. Z.; Ayad, S.; Weiss, J. B.: Isolation and native characterization of cysteine-rich collagens from bovine placental tissues and uterus and their relationship to types IV and V collagens. Biosci. Rep. *2:* 493–502 (1982).

2 Algire, G. H.; Chalkey, H. W.: Vascular reactions of normal and malignant tissues in vivo. I. Vascular reactions of mice to wounds and to normal neoplastic transplants. J. natn. Cancer Inst. *6:* 73–85 (1945).

3 Ayad, S. R.; Delinassios, J. G.: Esterase gene expression in Chinese hamster and mouse lymphoma hybrids isolated under nonselective pressure. Biochem. Genet. *12:* 147–161 (1974).

4 Azizkhan, R. G.; Azizkhan, J. C.; Zetter, B. R.; Folkman J.: Mast cell heparin stimulates migration of capillary endothelial cells in vitro. J. exp. Med. *152:* 931–944 (1980).

5 Brown, R. A.; Weiss, J. B.; Tomlinson, I. W.; Phillips, P.; Kumar, S.: Angiogenic factor from synovial fluid resembling that from tumours. Lancet *i:* 682–685 (1980).

6 Brown, R. A.; Tomlinson, I. W.; Hill, C. R.; Weiss, J. B.; Phillips, P.; Kumar, S.: Relationship of angiogenesis factor in synovial fluid to various joint diseases. Ann. rheum. Dis. *42:* 301–307 (1983).

7 D'Amore, P. A.; Glaser, B. M.; Brunson, S. K.; Patz, A.; Michels, R. G.; Fenselau, A.: Demonstration of a vasoproliferative activity from mammalian retina. Proc. natn. Acad. Sci. USA *78:* 3068–3072 (1981).

8 Elstow, S. F.; Schor, A.; Weiss, J. B.: Vasoproliferation activity in adult bovine retinas. Evidence for a low molecular weight angiogenesis factor. Ophthal. Res. (in press, 1984).

9 Epstein, E. H. Jr.: $(\alpha_1\ (III))_3$ Human skin collagen: release by pepsin digestion and preponderance in foetal life. J. biol. Chem. *249:* 3225–3231 (1974).

10 Fenselau, A. H.: Tumour and related angiogenesis factors; in Guroff (ed.), Growth and Maturation Factors (in press, 1984).

11 Fenselau, A. H.; Watt, S.; Mello, R. J.: Tumor angiogenic factor purification from the Walker 256 rat tumor. J. biol. Chem. *256:* 9605–9612 (1981).

12 Folkman, J.; Merler, E.; Abernathy, C.; Williams, G.: Isolation of a tumor factor responsible for angiogenesis J. exp. Med. *133:* 275–288 (1971).

13 Gisslow, V. T.; McBride, B. C.: A rapid sensitive collagenase assay. Ann. Biochem. *68:* 70–78 (1975).

14 Glaser, B. M.; D'Amore, P. A.; Michels, R. G.; Patz, A.; Fenselau, A.: Demonstration of vasoproliferative activity from mammalian retina. J. Cell Biol. *84:* 298–304 (1980).

15 Greenblatt, M.; Shubik, P.: Tumour angiogenesis: transfilter diffusion studies in the hamster by the transparent chamber technique. J. natn. Cancer Inst. *41:* 111–124 (1968).

16 Hill, C. R.; Weiss, J. B.: Preparation of low molecular weight angiogenesis factor from mouse lymphoma. (in press, 1984).
17 Keegan, A.; Hill, C. R.; Kumar, S.; Phillips, P.; Schor, A.; Weiss, J. B.: Purified tumour angiogensis factor enhances proliferation of capillary, but not aortic, endothelial cells in vitro. J. Cell Sci. *55:* 261–276 (1982).
18 Kissun, R. D.; Garner, A.: Vasoproliferative properties of normal and hypoxic retinal tissues. Br. J. Ophthal. *61:* 394–398 (1977).
19 Kissun, R. D.; Hill, C. R.; Garner, A.; Phillips, P.; Kumar, S.; Weiss, J. B.: A low molecular weight angiogenic factor in cat retina. Br. J. Ophthal. *66:* 165–169 (1982).
20 McAuslan, B. R.; Hoffman, H.: Endothelium stimulating factor from Walker carcinoma cells. Exp. Cell Res. *119:* 181–190 (1979).
21 Phillips, P.; Kumar, S.: Tumour angiogenesis factor (TAF) and its neutralisation by xenogeneic antiserum. Int. J. Cancer *23:* 82–88 (1979).
22 Ryan, T. J.; Stockley, A. T.: The chorioallantoic membrane of the embryonic chick as an assay for angiogenic factors. Biblthca anat., vol. 20, pp. 641–644 (Karger, Basel 1981).
23 Schor, A. M.; Schor, S. L.; Weiss, J. B.; Brown, R. A.; Kumar, S.; Phillips, P.: Stimulation by a low molecular weight angiogenic factor of capillary endothelial cells in culture. Br. J. Cancer *41:* 790–799 (1980).
24 Sellers, A.; Cartwright, E.; Murphy, G.; Reynolds, J. J.: Evidence that latent collagenases are enzyme inhibitor complexes. Biochemistry (Wash.) *163:* 303–307 (1977).
25 Stricklin, G. P.; Eisen, A. Z.; Bauer, E. A.; Jeffrey, J. J.: Human skin fibroblast collagenase: chemical properties of precursor and active forms. Biochem. *17:* 2331–2337 (1978).
26 Werb, Z.: Degradation of collagen; in Weiss, Jayson (eds.), Collagen in Health and Disease, pp. 121–134 (Churchill Livingstone, London 1982).
27 Weiss, J. B.; Brown, R. A.; Kumar, S.; Phillips, P.: An angiogenic factor isolated from tumours. A potent low molecular weight compound. Br. J. Cancer *40:* 493–496 (1979).
28 Weiss, J. B.; Hill, C. R.; Davis, R. J.; McLaughlin, B.; Sedowofia, K. A.; Brown, R. A.: Activation of a procollagenase by low molecular weight angiogenesis factor. Biosci. Rep. *3:* 171–177 (1983).
29 Weiss, J. B.; Hill, C. R.; McLaughlin, B.; Elstow, S. F.: Potentiating effect of heparin in the activation of procollagenase by a low molecular weight angiogenesis factor. FEBS Lett. *163:* 62–65 (1983).
30 Weiss, J. B.; Hill, C. R.; Schor, A.: Low molecular weight tumour angiogenesis factor. Biochem. Soc. Trans. (in press, 1984).

Dr. J. B. Weiss, Department of Rheumatology, University of Manchester, Medical School, Manchester M13 9PT (England)

Subject Index